BIOMIMETIC NANOENGINEERED MATERIALS FOR ADVANCED DRUG DELIVERY

BIOMIMETIC NANOENGINEERED MATERIALS FOR ADVANCED DRUG DELIVERY

Edited by:

AFEESH RAJAN UNNITHAN

ARATHYRAM RAMACHANDRA KURUP SASIKALA

CHAN HEE PARK

CHEOL SANG KIM

ELSEVIER

Elsevier
Radarweg 29, PO Box 211, 1000 AE Amsterdam, Netherlands
The Boulevard, Langford Lane, Kidlington, Oxford OX5 1GB, United Kingdom
50 Hampshire Street, 5th Floor, Cambridge, MA 02139, United States

Library of Congress Cataloging-in-Publication Data
A catalog record for this book is available from the Library of Congress

British Library Cataloguing-in-Publication Data
A catalogue record for this book is available from the British Library

ISBN: 978-0-12-814944-7

Publisher: Matthew Deans
Acquisition Editor: Sabrina Webber
Editorial Project Manager: Ali Afzal-Khan
Production Project Manager: Sojan P. Pazhayattil
Cover Designer: Mark Rogers

Typeset by SPi Global, India

Contents

Contributors

Ludwig Erik Aguilar
Bionanosystem Engineering, Chonbuk National University, Jeonju, Republic of Korea

Viswanathan A. Aparna
Amrita Centre for Nanosciences and Molecular Medicine, Amrita Institute of Medical Sciences and Research Centre, Amrita Vishwa Vidyapeetham, Kochi, India

Renu Geetha Bai
Department of Chemical Engineering, American University of Sharjah, Sharjah, United Arab Emirates

Raja Biswas
Amrita Centre for Nanosciences and Molecular Medicine, Amrita Institute of Medical Sciences and Research Centre, Amrita Vishwa Vidyapeetham, Kochi, India

Aneesh Chandrasekharan
Cancer Research Program, Rajiv Gandhi Centre for Biotechnology, Thiruvananthapuram, India

Pramod Darvin
Cancer Research Program, Rajiv Gandhi Centre for Biotechnology, Thiruvananthapuram, India

Amin GhavamiNejad
Advanced Pharmaceutics and Drug Delivery Laboratory, Leslie L. Dan Faculty of Pharmacy, University of Toronto, Toronto, ON, Canada

Ghaleb A. Husseini
Department of Chemical Engineering, American University of Sharjah, Sharjah, United Arab Emirates

R. Jayakumar
Amrita Centre for Nanosciences and Molecular Medicine, Amrita Institute of Medical Sciences and Research Centre, Amrita Vishwa Vidyapeetham, Kochi, India

Yong Yeon Jeong
Department of Radiology, Chonnam National University Hwasun Hospital, Hwasun, South Korea

Johnson V. John
Department of Chemical and Materials Engineering, University of Alberta, Edmonton, AB, Canada

Cheol Sang Kim
Department of Bionanosystem Engineering, Graduate School; Division of Mechanical Design Engineering, Chonbuk National University, Jeonju, Republic of Korea

Yeu Chun Kim
Department of Chemical and Biomolecular Engineering, Korea Advanced Institute of Science and Technology (KAIST), Daejeon, Republic of Korea

Joshua Lee
Department of Bionanosystem Engineering, Graduate School, Chonbuk National University, Jeonju, Republic of Korea

Brian Lu
Advanced Pharmaceutics and Drug Delivery Laboratory, Leslie L. Dan Faculty of Pharmacy, University of Toronto, Toronto, ON, Canada

Jason McCarthy
Masonic Medical Research Laboratory, Utica, NY, United States

Muthunarayanan Muthiah
Masonic Medical Research Laboratory, Utica, NY, United States

Chan Hee Park
Department of Bionanosystem Engineering, Graduate School; Division of Mechanical Design Engineering, Chonbuk National University, Jeonju, Republic of Korea

N. Sanoj Rejinold
Department of Chemical and Biomolecular Engineering, Korea Advanced Institute of Science and Technology (KAIST), Daejeon, Republic of Korea

T.R. Santhosh Kumar
Cancer Research Program, Rajiv Gandhi Centre for Biotechnology, Thiruvananthapuram, India

Arathyram Ramachandra Kurup Sasikala
Department of Bionanosystem Engineering, Graduate School; Division of Mechanical Design Engineering, Chonbuk National University, Jeonju, Republic of Korea; College of Medical and Dental Sciences, University of Birmingham, Birmingham, United Kingdom

S. Sowmya
Research Scientist, Amrita School of Dentistry, Amrita Vishwa Vidyapeetham, Kochi, India

Reju George Thomas
Department of Radiology, Chonnam National University Hwasun Hospital, Hwasun, South Korea

Afeesh Rajan Unnithan
Department of Bionanosystem Engineering, Graduate School; Division of Mechanical Design Engineering, Chonbuk National University, Jeonju, Republic of Korea; Institute of Translational Medicine, Chemical Engineering Department, University of Birmingham, Birmingham, United Kingdom

Xiao Yu Wu
Advanced Pharmaceutics and Drug Delivery Laboratory, Leslie L. Dan Faculty of Pharmacy, University of Toronto, Toronto, ON, Canada

About the editors

Dr. Afeesh Rajan Unnithan is currently working as a research fellow in Professor Alicia El Haj group at University of Birmingham, UK. Prior to this he was working as KRF (Korean Research Fellow) at Chonbuk National University, South Korea. Dr. Afeesh also worked as an assistant research professor at Chonbuk National University from 2013 to 2016. He completed his PhD in bionanosystem engineering from Chonbuk National University in 2013. Dr. Afeesh earned his MTech in nanomedicine from Amrita Center for Nanosciences and Molecular Medicine and BTech in biotechnology from Anna University, India. Dr. Afeesh's principal research interests are in the areas related to the preparation of nanobiomaterials, smart drug delivery systems, hyperthermic chemotherapy, bioactive nanostructured scaffolds, electrospinning, 3D bioprinting, etc. He has published around 49 peer-reviewed research articles in high impact factor journals with 1523 citations.

Dr. Arathyram Ramachandra Kurup Sasikala is currently working as a research fellow in Dr. Hanene Ali-Boucetta group at University of Birmingham, UK. Before coming to UK, she was working as an assistant research professor at Chonbuk National University, South Korea. She earned her PhD in bionanosystem engineering from Chonbuk National University in August 2016. Dr. Arathy completed her MSc and BSc in physics from University of Kerala, India. Dr. Arathy also worked as a junior research fellow in National Remote Sensing Centre, ISRO (Indian Space Research Organization), India. Her principal research interest lies in the development of multifunctional therapeutic nanosystems incorporating both magnetic nanomaterials and drugs for the synergistic cancer theranostics by combining hyperthermia, chemotherapy, and MRI (magnetic resonance imaging). Dr. Arathy published her research works in high impact journals such as Advanced Functional Materials, Nanoscale, Acta Biomaterialia, Scientific Reports, Journal of Materials Chemistry B, ACS applied materials and interfaces, Chemical Engineering Journal, etc. She has published around 15 peer-reviewed research articles in high impact factor journals with 443 citations.

Chan Hee Park is currently working as an associate professor in the Department of Bionanosystem Engineering at Chonbuk National University,

South Korea. He earned his PhD in bionanosystem engineering from Chonbuk National University in 2012. He worked for National Instruments (USA) from 2002 to 2009. Following that he worked as the team leader senior researcher at R&D Division, Chonbuk National University Automobile Parts & mold Technology Innovation Center, Jeonju. Professor Park's research interests are related to electrospun nanocomposite materials for drug delivery, microfluidics/electroanalytical biosensors, biodegradable metals, regenerative medicine, nanomedicine, etc. He published his research achievements in various peer-reviewed journals with 2124 citations.

Cheol-Sang Kim received his BS and MS degrees in mechanical engineering from Chonbuk National University in Korea in 1980 and 1982, respectively. He then earned his PhD in material science from Universite de Louis Pasteur in Strasbourg, France in 1988. Dr. Kim is currently a professor at the Division of Mechanical Engineering and Dean of College of Engineering at Chonbuk National University in Korea. Prior to joining Chonbuk National University, he has spent 2 years at the Department of Bioengineering at University of Pennsylvania (USA) as a postdoctoral fellow. He worked for 5 years from 1997 as a Head of Institute of Biomedical Engineering, Solco Surgical Instruments Co., Ltd. and Institute of Interventional Medicine, M.I. Tech Co., Ltd., Korea. Dr. Kim's research interests are in the area of biomaterials for hard tissue replacements, drug delivery, design and analysis of implants and artificial organs, and anti-biofouling technology. Dr. Kim owns numerous publications in high impact factor journals with 2426 citations.

CHAPTER 1

Introduction to smart drug delivery systems

Pramod Darvin, Aneesh Chandrasekharan, T.R. Santhosh Kumar
Cancer Research Program, Rajiv Gandhi Centre for Biotechnology, Thiruvananthapuram, India

Contents

1.1 Overview

The use of nanotechnology and their applications have revolutionized in this century. The demands range from the needs of daily life to the unmet demand for managing chronic diseases in medical field. In the medical field, tremendous applications are being utilized in different aspects of clinical management. The use of nanotechnology in pharmacology and drug delivery system enabled the reduction of nonspecific biodistribution and uncontrolled drug release. Focus to address this issue and achieve drug release at a precise location with more control resulted in the development of smart drug delivery systems (SDDSs). These SMART systems can reach the précised site where the drug is intended to release, and can also release the drug in response to specific stimulations. This ability makes them intelligent systems, capable of self-regulation, integrated sensing, monitoring, and activation by the environment as well as stimuli.

Biomimetic Nanoengineered Materials for Advanced Drug Delivery
https://doi.org/10.1016/B978-0-12-814944-7.00001-1

1.2 Stimuli-responsive SDDS

SDDS were designed to release the payloads to the specific site in a more controlled manner so as to increase the therapeutic efficiency and reduce the adverse effects. Stimuli-responsive systems respond to both external and internal stimuli. The external stimuli include the magnetic, electric, thermal, and ultrasonic form of energy. The internal stimuli are the redox status, pH, and temperature of the system, biochemical factors like specific enzymes, urea, glucose, morphine responsive system, and specific body conditions such as cancer and inflammation responsive system.

1.2.1 Temperature-responsive SDDS

Thermosensitive polymers are designed to retain their payloads around the physiological temperature (37°C) and the drugs will be released rapidly when the temperature rises more than 40–45°C. These systems are suitable for the controlled delivery of drugs and genes. Thermosensitive polymers undergo sol-gel transitions in response to temperature and minute changes in physiological temperature reflect the solubility changes. In the clinical conditions like inflammation and malignancies, the temperature of pathologic tissues rises compared to the normal tissues. The temperature-responsive SDDS can recognize this change in temperature and are activated. The advantage of using temperature-responsive SDDS is that the stimuli can be applied either as internal stimuli or as external stimuli. Induced temperature generation at the malignant site helps also to release the drug into the tumor microenvironment. For example, the poly(*N*-isopropyl acrylamide) (PNIPAAm) is a thermosensitive polymer with a lower critical solution temperature of 32°C which is adjusted to physiological temperature using additives or surfactants. When the temperature rises above 27°C, the solution becomes cloudy and forms a gel. Multiple thermo-responsive nanocarrier structures like hydrogels, microbeads, polymeric nanotubes, micelles, core-shell thermo-responsive nanoparticles, and layer-by-layer assembled nanocapsule have been developed using thermosensitive polymers and are being used as SDDS. Thermosensitive polymers commonly in use are PNIPAAm, poly(*N*,*N*-diethylacrylamide), poly(*N*-vinylalkylamide), poly(*N*-vinylcaprolactam), tetronics, pluronics, phosphazene derivative, polysaccharide derivatives, etc. [1–3].

Hydrogels due to their high-water content undergo a variety of changes upon external stimulation. They can undergo swelling, shrinking, and sol-gel phase transition. Cellulose derivatives, PNIPAAm, poly(lactic-co-glycolic acid) (PLGA), polycaprolactone, and polyurethane amide are some of the widely used hydrogels. PNIPAAm is one of the highly studied hydrogels for drug delivery purposes. They have high thermal reversibility and good swelling ratio. Copolymerization and grafting of PNIPAAm with polyethylene glycol (PEG) and polyethylene oxide (PEO) have established a route to obtain new hydrogels. PNIPAAm-derived hydrogels face potential drawbacks like increased cytotoxicity and low biodegradability. Because of the applicability of the hydrogels, the modification is made to the backbone of PNIPAAm copolymers to increase the biodegradability. The incorporation of biodegradable segments using cleavable bonds has improved biodegradation and biocompatibility.

Hydrogels having transition temperature near physiological temperature are used as nonsurgical drug depots and for sustained delivery of drugs and proteins to susceptible living cells. Transition temperature can be readily tuned by copolymerization conditions and by varying the content of repeating units in the copolymer, for example, copolymerization of acrylic acid (AAc) and NIPAAm.

Burst drug-release kinetics may be sometimes observed in the case of hydrogels and porous capsules with low strength that is easily eroded. Additionally, hydrogels can suffer from poor encapsulation of hydrophobic drugs due to the hydrophilic nature of the hydrogel. Poloxamer-based hydrogels and PLGA-PED-PLGA hydrogels are examples of encapsulated hydrogels with the sustained drug-release profile.

Polymeric micelles with specific core/shell architecture entrap hydrophobic drugs in the inner core surrounded by an outer shell of hydrophilic polymers. Micelle-based drug delivery system is used for anticancer drug delivery due to its ease of administration, reduced system toxicity, high payload, and therapeutic efficiency. Thermostability depends on the cloud point of the thermosensitive block of the copolymer forming the micelle. The reversible phase transition is a mechanism generally employed by the thermosensitive micelles.

Based on the application, the temperature-responsive delivery systems are designed. Multifunctional and dual stimuli-responsive systems have been developed to release the cargo in response to more than one stimulus. For example, temperature-pH-sensitive systems, temperature-pH reduction-sensitive polymeric micelle, and macromolecule-sensitive nanogels like glucose-pH-thermo-responsive nanogels have been developed. Thermo-responsive nanostructures have advantages like low toxicity and target specificity.

1.2.2 Redox-responsive drug delivery system

Redox-sensitive drug delivery system has received great attention for its close connection with many diseases and is extensively studied. Moreover, the redox-sensitive delivery system has the advantage of intracellular drug release. The variation in the redox status of tissues is exploited in the designing of this system. Intracellular glutathione-dependent systems are the major redox-responsive SDDS. The concentration of GSH (glutathione) varies between normal and malignant cells. It ranges from 2 to 20 μM in blood and normal extracellular matrices and from 2 to 10 μM in cancer cells [4]. Other disease conditions like inflammation, the ROS (reactive oxygen species) level increases 10- to 100-fold compared with normal counterparts.

Drug delivery systems with disulfide linkage and diselenide linkage are the two major reduction-based nano-drug delivery systems [4, 5]. The disulfide linkages can be broken with high concentrations of GSH, resulting in a quick release of the cargoes. These disulfide bonds are mainly used as linkers and cross-linking agents [4, 6]. As a linker, the disulfide linkage connects the polymer backbone with drugs or other cargoes. In the backbone, the disulfide linkage breaks rapidly with high concentrations of GSH in the surrounding environment. This makes the polymer unstable under certain situations. In the backbone, cystamine [7], cystine [8], *N*-succinimidyl-3-(2-pyridyldithiol) propionate (SPDP) [9], disulfide-based dimethacrylate (DSDMA) [10], etc., are used as disulfide-containing fragments. The disulfide linkages present in the side chains are intended to support the backbone. The polymers with disulfide linkages in the side chains are easy to modify and can have high cargo-carrying capacity and would be more stable than the drug delivery systems with disulfide linkages in the backbone. This linker disulfide bonds can enrich the nanocarrier by attaching multiple genes, drugs, and targeting molecules to their surface. These bonds can link two different moieties with different functions and can act as cross-linking agents.

Redox-sensitive delivery system containing diselenide bonds are also sensitive to the reduction and show similar properties like disulfide linkages. The selenide bond shows bond energy (diselenide bond, $172\,kJ\,mol^{-1}$ and carbon-selenium bond, $244\,kJ\,mol^{-1}$) lower than disulfide bonds, making them suitable for developing more sensitive delivery systems. The major challenge using diselenide bond containing polymer is the difficulty in the incorporation of selenide bond.

Oxidation-responsive drug delivery systems is another major group of drug delivery systems in redox-sensitive systems and mainly rely on the

ROS, mainly H_2O_2 and –OH radicals. ROS are omnipresent in tissues and are associated with pathological conditions like arteriosclerosis, heart and nerve injuries and inflammations. These groups of drug delivery systems include the sulfide-containing systems such as poly(propylene sulfide), selenium-containing systems, ferrocene-containing systems, boronic ester groups, etc. [11–14].

1.2.3 pH-responsive systems

In most of the applications related to diseases, the drugs need to be delivered to specific sites in order to decipher maximum efficacy with minimum toxicity. pH-responsive systems are a better choice for this and are commonly used SDDS due to the change in pH among the organs, organelles, and tissues. This system also confers precise delivery options and predetermined release rates. All pH-responsive polymers used contain a pendant acidic (carbonic and sulfonic acids) or basic (ammonium salt) groups as a proton acceptor or donor. These carriers can trace delicate changes in the pH of inflammatory or other malignant sites. Generally, the tumor microenvironment has less pH (7.0) compared to their normal counterparts (pH 7.4). The change in environmental pH causes the ionization of pendant acidic or basic groups resulting in the cross-linking and changes in the swelling properties of the polymer. The precision and accuracy of pH-responsive systems are increased by combining them with other stimuli such as temperature or redox stimuli as multifactor-responsive drug-release system [2].

The nature of the drug release is designed by the pH of the solution to which the polymer is exposed. The alteration in pH causes the polymer to shrink or swell. For example, polyacidic polymers will shrink at low pH and swell at high pH because of the protonation of acidic groups. In the case of polybasic polymers, ionization of basic groups increases with increase in pH. AAc and cellulose derivatives are the most widely used pH-sensitive polymers. More than two phases can occur in copolymers with randomly distributed positive and negative charged groups. In these copolymers, the interaction of segments with each other occurs through electrostatic interactions and hydrogen bonding. The existence of these interactions helps the polymer to shrink and swell according to the changes in the environmental conditions and adopt a stable conformation.

Polyacrylic acid (PAAc) and polymethacrylic acid (PMAAc) are the examples of the polyacidic pH-responsive system. PMAAc grafted with PEG show unique pH sensitivity. Hydrogels made from PAAc or PMAAc can be employed for targeted release of drugs at neutral pH. For colon-specific drug delivery, polyanions cross-linked with azoaromatic cross-linkers were used. Sulfonamide contains polyacid polymers with pK_a in the range 3–11 and the hydrogen atom of the amide nitrogen is readily ionized to form polyacids. The amino group is protonated at high pH and positively neutralized and ionized at low pH controlling the swelling and shrinking nature of hydrogels. Polybases with the amino group are the most representative polybasic group. Poly(*N*,*N*-dimethylaminoethylmethacrylate) (PDEAEMA) has been the most frequently used pH-responsive polymeric base. Polycationic hydrogels show minimum swelling in the neutral pH, minimizing the release of drug cargo. This property of polycationic hydrogels has been used to develop carriers for foul-tasting drugs, to prevent them from releasing into the neutral pH environment of the mouth.

The pH responsiveness of polyelectrolyte hydrogels can be modulated by using neutral comonomers. Various comonomers show different hydrophobicity and hence different pH responsiveness. 2-Hydroxyethyl methacrylate, methyl methacrylate maleic anhydride, etc., are the examples of comonomers used in polyelectrolyte hydrogels.

pH-sensitive bioerodible polymers use enzyme-substrate reaction. An enzyme-substrate reaction produces a change in pH and this change in pH is utilized for the erosion of a pH-sensitive polymer containing a drug cargo. In addition to the bioerodible polymers, rDNA methods and protein engineering were employed to develop synthetic proteins that can undergo reversible gelation in response to pH changes. pH-sensitive nanoparticles enhance drug availability through efficient absorption mechanisms. Polymers such as hydroxypropylmethylcellulose phthalate and cellulose acetate phthalate were used to encapsulate the drug load. Ionic gelation, solvent evaporation, solvent diffusion, supercritical fluid technology, salting out, polymerization, etc., can be employed to produce pH-sensitive polymeric nanoparticles.

1.2.4 Enzyme-responsive drug delivery systems

Enzyme activated drug delivery system is a new class of delivery system where the drug release is controlled by enzyme reactions. Several nanomaterials and polymers are being used for the development of enzyme-responsive drug delivery systems. This system has the advantage of a high degree of specificity. Most of the physiological and metabolic processes are related with enzymes such as glycosidase, lipase, phospholipase, and protease. These enzymes are exploited to achieve enzyme-mediated drug cargo release through biocatalytic action at inflammatory or malignant sites. The nanocarriers carrying payload attached with them through encapsulation or covalent bonding will release the drugs at the target sites through site-specific enzymatic cleavage. Multiple enzymes are used to trigger the drug-release system [15]. Different classes of protease are involved in various physiological and pathophysiological conditions like wound healing, tumor invasion, and tissue remodeling. Protein kinase C-alpha, uPA (urokinase-type plasminogen activator), and MMPs (matrix metalloproteinases) are the commonly used protease enzymes in the controlled drug release [16–18]. The major challenge in enzyme-responsive drug delivery system is to precisely control the initial response time of the system. Phospholipases are another important group of enzymes involved in infections, neurodegenerations, malignancies, and inflammations. Phospholipase A2 (PLA2) enzyme is utilized to release drugs or expose target ligands from liposome or small unilamellar vesicle (SUV)-mediated drug delivery systems [19, 20]. Glucose oxidase of oxidoreductase family, involved in glucose metabolism, has its application in the diagnosis and drug delivery in response to glucose levels [21].

1.2.5 Magnetic-responsive drug delivery systems

Magnetic-responsive drug delivery systems provide a noninvasive approach to spatiotemporal control of the carriers to the targets. This helps the system to release the payloads under programmable exposure of external magnetic fields. The most commonly used core/shell magnetic nanoparticle exhibits a variety of unique magnetic properties. The complex consists of the drug and pharmaceutically stable ferromagnetic carrier. Based on the fabrication method the surface-to-volume ratio of MNPs (magnetic nanoparticles) can be adjusted. Increased surface-to-volume ratio provides numerous active sites for the conjugation of biomolecules. This allows precision in design and engineering to achieve their smart functions like elevated blood half-life, release site specificity, and increased delivery by applying a localized external magnetic field [22].

Furthermore, when these nano-scaled MNPs were encapsulated in colloidal carriers, such as micelles, liposomes (Magneto liposomes), and solid nanoparticles, they become more sensitive to an external magnetic field. These magnetic nanoparticles also enable the engineering of multi-stimuli-responsive drug delivery systems which can act on pH, redox, and enzymes [23, 24]. For example, cystamine-modified gelatin encapsulated nanocarrier containing doxorubicin as the payload showed increased drug target specificity, EPR, and good biocompatibility [25].

1.3 Conclusion

The development of smart drug delivery system offers great potential to engineer pathology specific drug design and delivery approaches suiting to the clinical requirement. However, the field further requires increased attention to address the challenges of clinical acceptability and clinical translations. The development of more nano-drug carriers is underway to even increase the smaller physicochemical changes in pathologic and normal tissues. More combinations and multifactorial release systems are attractive because of their advance in selectivity and specificity thereby increasing the overall efficacy of the drugs.

References

[1] H. Priya James, et al., Smart polymers for the controlled delivery of drugs—a concise overview, Acta Pharm. Sin. B 4 (2) (2014) 120–127.
[2] Y. Qiu, K. Park, Environment-sensitive hydrogels for drug delivery, Adv. Drug Deliv. Rev. 53 (3) (2001) 321–339.
[3] S. Choi, M. Baudys, S.W. Kim, Control of blood glucose by novel GLP-1 delivery using biodegradable triblock copolymer of PLGA-PEG-PLGA in type 2 diabetic rats, Pharm. Res. 21 (5) (2004) 827–831.
[4] M. Huo, et al., Redox-responsive polymers for drug delivery: from molecular design to applications, Polym. Chem. 5 (5) (2014) 1519–1528.
[5] T. Thambi, et al., Bioreducible polymersomes for intracellular dual-drug delivery, J. Mater. Chem. 22 (41) (2012) 22028–22036.
[6] X. Guo, et al., Advances in redox-responsive drug delivery systems of tumor microenvironment, J. Nanobiotechnol. 16 (1) (2018) 74.
[7] Y. Li, et al., Synthesis of reversible shell cross-linked micelles for controlled release of bioactive agents, Macromolecules 39 (8) (2006) 2726–2728.
[8] J. Wu, et al., Hydrophobic cysteine poly(disulfide)-based redox-hypersensitive nanoparticle platform for cancer theranostics, Angew. Chem. Int. Ed. Eng. 54 (32) (2015) 9218–9223.
[9] Y. Vachutinsky, et al., Antiangiogenic gene therapy of experimental pancreatic tumor by sFlt-1 plasmid DNA carried by RGD-modified crosslinked polyplex micelles, J. Control. Release 149 (1) (2011) 51–57.
[10] L. Zhang, et al., Degradable disulfide core-cross-linked micelles as a drug delivery system prepared from vinyl functionalized nucleosides via the RAFT process, Biomacromolecules 9 (11) (2008) 3321–3331.
[11] A. Napoli, et al., Oxidation-responsive polymeric vesicles, Nat. Mater. 3 (2004) 183.
[12] N. Ma, et al., Selenium-containing block copolymers and their oxidation-responsive aggregates, Polym. Chem. 1 (10) (2010) 1609–1614.

[13] X. Sui, et al., Redox active gels: synthesis, structures and applications, J. Mater. Chem. B 1 (12) (2013) 1658–1672.
[14] C.-C. Song, et al., Oxidation-accelerated hydrolysis of the ortho ester-containing acid-labile polymers, ACS Macro Lett. 2 (3) (2013) 273–277.
[15] Q. Hu, P.S. Katti, Z. Gu, Enzyme-responsive nanomaterials for controlled drug delivery, Nanoscale 6 (21) (2014) 12273–12286.
[16] B. Law, R. Weissleder, C.-H. Tung, Peptide-based biomaterials for protease-enhanced drug delivery, Biomacromolecules 7 (4) (2006) 1261–1265.
[17] T. Jiang, et al., Tumor imaging by means of proteolytic activation of cell-penetrating peptides, Proc. Natl. Acad. Sci. U. S. A. 101 (51) (2004) 17867.
[18] E.S. Olson, et al., Activatable cell penetrating peptides linked to nanoparticles as dual probes for in vivo fluorescence and MR imaging of proteases, Proc. Natl. Acad. Sci. 107 (9) (2010) 4311.
[19] L. Linderoth, et al., Drug delivery by an enzyme-mediated cyclization of a lipid prodrug with unique bilayer-formation properties, Angew. Chem. Int. Ed. 48 (10) (2009) 1823–1826.
[20] T.L. Andresen, et al., Enzymatic release of antitumor ether lipids by specific phospholipase A2 activation of liposome-forming prodrugs, J. Med. Chem. 47 (7) (2004) 1694–1703.
[21] Z. Gu, et al., Glucose-responsive microgels integrated with enzyme nanocapsules for closed-loop insulin delivery, ACS Nano 7 (8) (2013) 6758–6766.
[22] S. Nappini, et al., Magnetic field responsive drug release from magnetoliposomes in biological fluids, J. Mater. Chem. B 4 (4) (2016) 716–725.
[23] Z. Qi, et al., A multi-controlled drug delivery system based on magnetic mesoporous Fe3O4 nanopaticles and a phase change material for cancer thermo-chemotherapy, Nanotechnology 28 (40) (2017) 405101.
[24] B. Mu, et al., Magnetic-targeted pH-responsive drug delivery system via layer-by-layer self-assembly of polyelectrolytes onto drug-containing emulsion droplets and its controlled release, J. Polym. Sci. A Polym. Chem. 49 (9) (2011) 1969–1976.
[25] A. Qi, et al., Gelatin-encapsulated magnetic nanoparticles for pH, redox, and enzyme multiple stimuli-responsive drug delivery and magnetic resonance imaging, J. Biomed. Nanotechnol. 13 (11) (2017) 1386–1397.

CHAPTER 2

Nanofiber-based anticancer drug delivery platform

Arathyram Ramachandra Kurup Sasikala[a,b,c], Afeesh Rajan Unnithan[a,b,d], Chan Hee Park[a,b], Cheol Sang Kim[a,b]
[a]Department of Bionanosystem Engineering, Graduate School, Chonbuk National University, Jeonju, Republic of Korea
[b]Division of Mechanical Design Engineering, Chonbuk National University, Jeonju, Republic of Korea
[c]College of Medical and Dental Sciences, University of Birmingham, Birmingham, United Kingdom
[d]Institute of Translational Medicine, Chemical Engineering Department, University of Birmingham, Birmingham, United Kingdom

Contents

2.1 Introduction

Cancer is the second foremost causes of death on globally just behind the cardiovascular disease demanding vital attention to significantly reduce the global cancer burden. [1] Cancer is a highly heterogeneous and multifaceted disease. In cancer, the cells undergo mutation in their genetic material and result in uncontrolled growth. Therefore the systemic delivery of antineoplastic agents such as chemotherapy or antiangiogenic drugs, radiotherapy, and surgery have been the traditional methods of treating cancer for a long time. Though it is successful for some types of cancer, the main problem associated with such treatment is the poor targetability and highly invasive nature [2]. Hence an extensive research is going on in search of an improved or novel treatment to combat this deadly disease by developing new drug delivery platforms that have extreme tumor targetability and minimal toxicity toward normal cells.

Biomimetic Nanoengineered Materials for Advanced Drug Delivery
https://doi.org/10.1016/B978-0-12-814944-7.00002-3

With the development of nanotechnology, cancer nanomedicine, the application of nanomedicine for the treatment of cancer, has seen enormous progress in recent years [3, 4]. To date several types of organic and inorganic nanomaterials in the form of nanoparticles, microparticles, liposomes, dendrimers, nanotubes, etc., were engineered to function as effective cancer theranostic agents [5–9]. The main advantages of the nanoengineered materials include the capability to increase the solubility of water-insoluble drugs, long-term blood circulation, and the ease of functionalization [10–12]. However, even with these technologies, there still remain many shortcomings associated with these nanocarriers. The nanocarriers could tide over numerous barriers starting from the injection site to the target cell, such as mucosal barriers and nonspecific uptake [13, 14]. The intravenously delivered nanosystems display overreliance on the enhanced permeation and retention (EPR) effect to deliver the nanocarriers into the tumor [15] as well as they suffer rapid clearance from the blood stream with subsequent overaccumulation in nontarget organs [16, 17].

The limited success and toxicity associated with current systemic administration of drugs has motivated researchers to search for "local therapy systems (LTSs)," a direct approach to deliver agents for cancer therapy. LTS can be considered as a promising alternative to systemic delivery by supplying the drugs directly to the targeted site of interest via an implantable system [18] which results not only in higher therapeutic efficacy and lower toxicity [19, 20] but also reduces the need for repeated chemotherapy application [21]. The LTSs are typically designed to be implanted immediately after a tumor resecting surgery, which spare the requirement for an additional surgery to place the therapeutic material in the patient [18] and enable the on-site delivery of drug by sidestepping the harsh environment and longer journey that the drug has to take to reach the site of interest when delivered systematically. Thus LTS can be regarded as an intelligent strategy to improve the quality of life and to enhance the patient compliance by repurposing promising drugs for cancer treatment.

Recently, electrospun polymer fibers have gained a great attention as implantable anticancer drug delivery devices due to their unique architectural features and robust drug loading capability with very encouraging preliminary results [22–27]. Nanofibers can be regarded as a versatile one-dimensional nanomaterial for a wide range of applications not only in the field of research but also in commercial arenas [28]. Among the broad spectrum of intriguing nanomaterials with great potential applications, nanofibers stand preeminent due to their outstanding physical and chemical properties [29]. The main advantages of the nanofibers are their exceptionally high surface area-to-volume ratio, unique porosity with interconnected pore, and biomimetic extracellular material (ECM)-like structure make them an attractive candidate for many advanced applications such as musculoskeletal tissue engineering (including bone, cartilage, ligament, and skeletal muscle), skin engineering, vascular tissue engineering, neural tissue engineering, and as carriers for controlled delivery of drugs, proteins, and DNA [28, 29]. This chapter aims at summarizing and discussing the state-of-the-art nanofiber-based anticancer drug delivery platforms to provide a comprehensive and significant scientific base supporting future strategic direction in cancer therapies.

2.2 State-of-the-art nanofiber fabrication

Electrospinning is considered as one of the simplest top-down approaches used to generate nanofibers [30]. Electrospinning can be performed using a wide variety of materials such as polymers, sol-gel, suspension, and melts [31]. The accessibility of a wide range of natural and synthetic biomaterials has extended the scope of developing nanofibrous scaffolds [29]. By varying the material selection and chemical compositions, the nanofibers obtain tunable material properties such as mechanical strength, biodegradability, wettability, and so on [28, 32].

The conventional electrospinning setup is shown in Fig. 2.1. In general, the electrospinning setup consists of three basic parts: (1) high-voltage supply, (2) a needle connected to syringe and pump, and (3) the collector, which is usually grounded or negatively biased. During electrospinning process high voltage is applied to create an electrically charged jet of polymer solution. When this charged polymer solution is pumped out through the spinneret, it forms a spherical droplet due to the internment of surface tension. Mutual charge repulsion and the contraction of the surface charges to the counter electrode cause a force directly opposite to the surface tension. If the intensity of the electric field escalates, the spherical surface of the polymer at the tip of the capillary tube extends to form a conical shape termed as the Tayler cone [33]. As the electrostatic force acting on the surface of the liquid overcomes the surface tension, the charge liquid jet is pulled from the Taylor cone and ejected directly toward the target collector. The charged jet undergoes instability and elongation process which cause the polymer fiber to become long and thin. As the solvent evaporates the polymer fibers were formed on the collector. The required modifications to the conventional

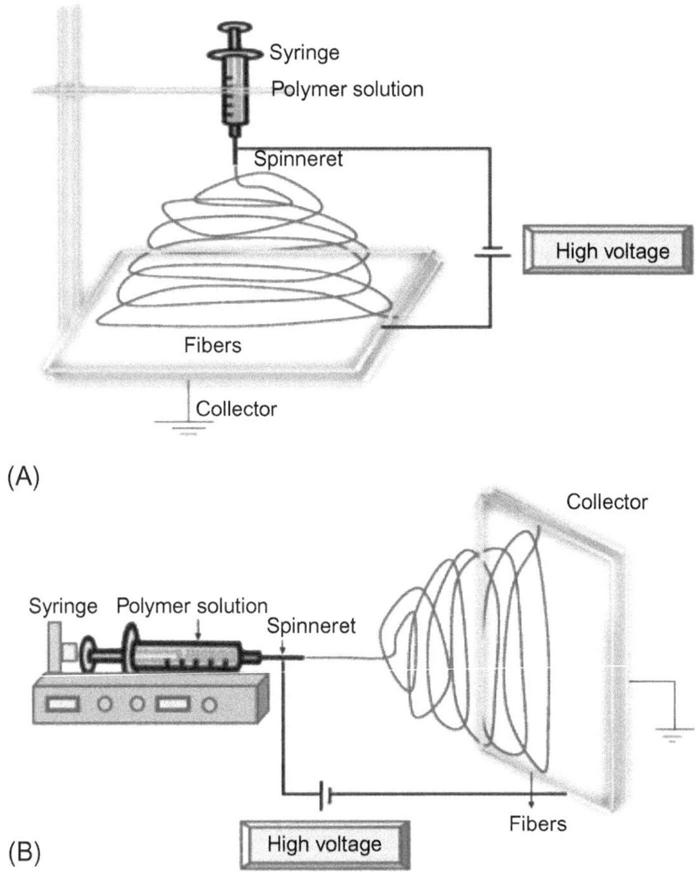

Fig. 2.1 Schematic representation of electrospinning apparatus (A) typical vertical system and (B) horizontal system. *(Reproduced with permission from E.A. Ho, P.L. Soo, C. Allen, M. Piquette-Miller, Impact of intraperitoneal, sustained delivery of paclitaxel on the expression of P-glycoprotein in ovarian tumors, J. Control. Release 117 (2007) 20–27, Elsevier.)*

electrospinning setup have been introduced for improving the quality and amendment of the nanofiber properties according to the requirement of the proposed applications [28] (Fig. 2.1).

The success of the electrospinning is influenced by various parameters such as solution parameters and environmental parameters. Solution parameters include viscosity, conductivity, surface tension, and the elasticity of the polymer solution. Various environmental factors such as electric potential, tip to collector distance, humidity, solution temperature, hydrostatic pressure on the solution, the air velocity of the system, etc., can also act as governing variables for electrospinning process [34]. So by fine-tuning the variable parameters we can control the geometry and bulk properties of nanofibers. In addition to that electrospinning also afford various methods for incorporating drug and biomolecules depending on the range of application. Various methods have been adopted to successfully incorporate active molecules into the nanofibers through the surface modification techniques along with modified electrospinning setup by varying the nozzle configuration to coaxial, multi-jet, multilayer, emulsion electrospinning, etc. [35–37] Each reported method has its own unique characteristics and the selection of appropriate method depends on the need of the experiments such as controlled initial burst release, high drug loading, long time controlled release, prolong circulation, etc. Among the mentioned techniques, co-electrospinning is considered to be the cost effective and simple method in which the polymer solution and the selected biomolecule or drug can be premixed before electrospinning. Moreover, the co-electrospinning ensures high drug loading, homogenous mixing of drug, and good initial burst release [38].

2.3 Implantable electrospun nanofibers as local therapy systems for cancer therapy

As described in the introduction, the currently available conventional treatments for cancer exhibit marginal efficacy with serious side effects. In order to overcome these side effects, targeted drug delivery system finds promising application. A targeted drug delivery system is exploited to selectively target and release the therapeutics only to the site of action and not to the nontarget organs or tissues [39] thus limiting the possible side effects to normal cells. There are two main targeted drug delivery methods which include nanoparticle-based method and localized drug delivery methods using implantable materials [40]. Electrospun nanofiber-based drug delivery system comes under the second category. Recent literature has described the combination stimuli-responsive smart drug delivery mechanism for effective anticancer drug delivery application rather than the conventional diffusion-based drug delivery mechanism [24, 41].

2.3.1 Electrospun nanofibers for chemotherapeutic delivery

Chemotherapy is a well-known modality to treat cancer, in which chemical substances that act electively on cells in mitosis and antimitotic agents were chosen to finally destroy the cancer cells [42]. Though chemotherapy is an effective treatment against cancer but undesirable chemotherapy reactions and the development of resistance to drugs which results in multidrug resistance are the major hurdles in chemotherapy [43]. In order to avoid these shortcomings, targeted therapy has been emanated as a promising treatment due to its specificity toward cancer cells and inactiveness toward off-target cells [44]. Targeted therapy targets and delivers drugs to specific genes or proteins that are specific to cancer cells or the tissue environment that promotes cancer growth [43].

Yang et al. [45] have developed an implantable active targeting micelle for anticancer therapy and incorporated the micelle into an implantable core shell polymeric nanofiber matrix for the systemic delivery of chemotherapeutics into the tumor region. The nano-micelles were prepared from a folate conjugated poly(ε-caprolactone)-poly(ethylene glycol) (FA-PCL-PEG) copolymer and encapsulated the anticancer drug doxorubicin (DOX) by self-assembling to act as an anticancer platform and finally the micelles were trapped in the core region of the core shell polymeric nanofibers comprising poly(vinyl alcohol) (PVA) as the core and gelatin (GEL) as the shell. They have tested the efficacy of the system by in vivo tumor model by implanting the nanofiber device subcutaneously near the artificial solid tumors and active targeting micelles were subsequently released from the core-shell nanofiber through degradation, which rapidly accumulated around the tumor tissue via interstitial transport and well-known EPR effect and specifically internalized by tumor cells via FR-mediated endocytosis. Thus they proved that combination of an active targeting micellar system and an implantable system is a highly efficient anticancer therapy which not only eliminates the repeated administration of therapeutics via intravenous injection of micelles but also reduces the drug dose as well as the side effects of chemotherapeutic agents.

Zhang et al. [46] reported an implantable prodrug micelle co-loaded nanofiber for local cancer treatment. For that, they first prepared the Pt(IV) prodrug-backboned micelle, which is further co-electrospun with dichloroacetate (DCA), a metabolic modulator, into biocompatible PVA nanofibers. The main advantages of such implantable prodrug micelle co-loaded nanofiber are that they dissolve after implantation at the tumor site and enable the fast release of DCA and Pt(IV) micelle to enable a synergistic anticancer efficacy by exhibiting different cell death mechanism as shown in Fig. 2.2.

Ranganath et al. [47] developed a paclitaxel (PTX)-loaded disc- and sheet-shaped biodegradable microfiber implants for malignant glioma treatment. Both microfiber sheets (MFSs) and microfiber discs (MFDs) were

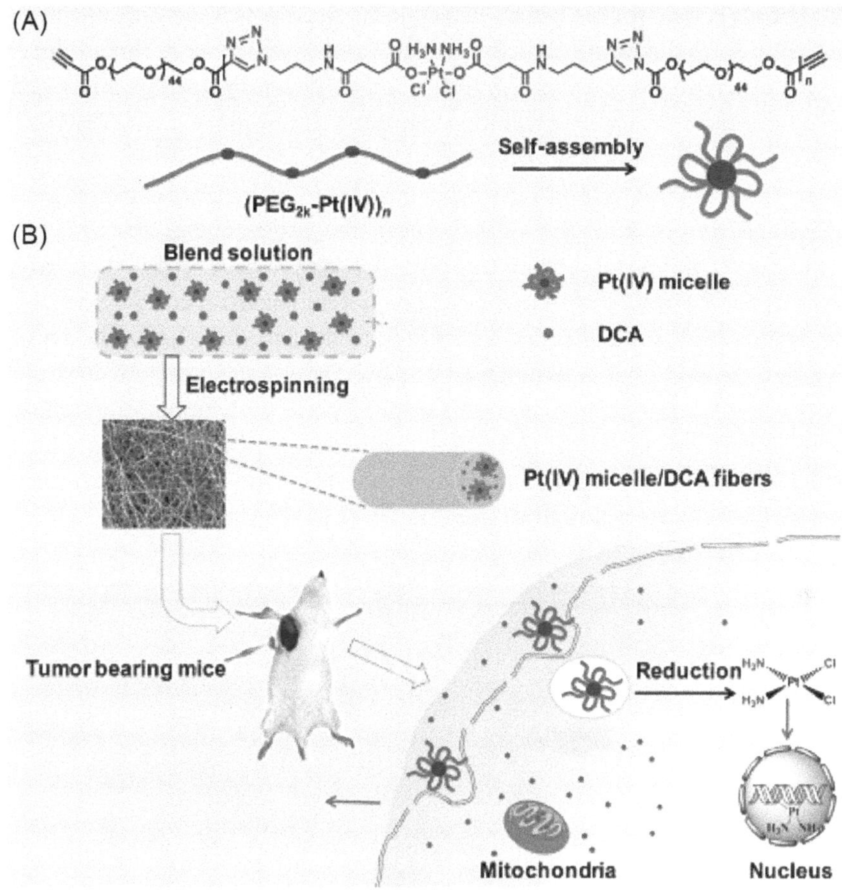

Fig. 2.2 Schematic illustration showing (A) the structure and self-assembly of Pt(IV) prodrug-backboned micelle and (B) the fabrication and the role of implantable prodrug micelle co-loaded nanofiber for local combination cancer therapy. *(Reproduced with permission from Z.Y. Zhang, Y.J. Wu, G.Z. Kuang, S. Liu, D.F. Zhou, X.S. Chen, X.B. Jing, Y.B. Huang, Pt(IV) prodrug-backboned micelle and DCA loaded nanofibers for enhanced local cancer treatment. J. Mater. Chem. B 5 (2017) 2115–2125. Royal Society of Chemistry.)*

fabricated by electrospinning poly-(D,L-lactide-co-glycolide) (PLGA) 85:15 copolymer and PLGA 50:50 copolymer was used to fabricate submicrofiber sheets (SFSs) and submicrofiber discs (SFDs) to profit different drug release properties. They observed that all the dosage forms exhibited a sustained PTX release over 80 days in vitro with a small initial burst. Apoptosis study confirmed the greater efficacy of nanofiber system for the sustained release of PTX compared to the critical Taxol administration. The main advantage of the fibrous system for the postsurgical glioma treatment is the implantability of the fibers in the tumor resection cavity along with the large surface area-to-volume ratio for the effective drug release. The animal study confirmed the antitumor efficacy by exhibiting inhibited tumor growth of 75%, 78%, 69%, and 71% for MFD, SFD, MFS, and SFS groups over placebo controlled groups after 24 days of tumor growth and concluding the crucial role of such fiber-based implants for the local chemotherapy of brain tumors.

Liu et al. [48] developed doxorubicin-loaded polylactide (PLA) electrospun nanofibers as a local chemotherapy system against secondary hepatic carcinoma (SHCC) and tested its efficacy in in vivo models. As shown in Fig. 2.3, the DOX-PLA nanofiber was placed after exploratory laparotomy by covering the surface of the nodular SHCC and by wrapping the whole liver bearing diffuse SHCC. After the treatment, the nodular SHCC exhibited a significant tumor reduction and the median survival time of the mice bearing DSHCC was found to increase from 14 days to 38 days.

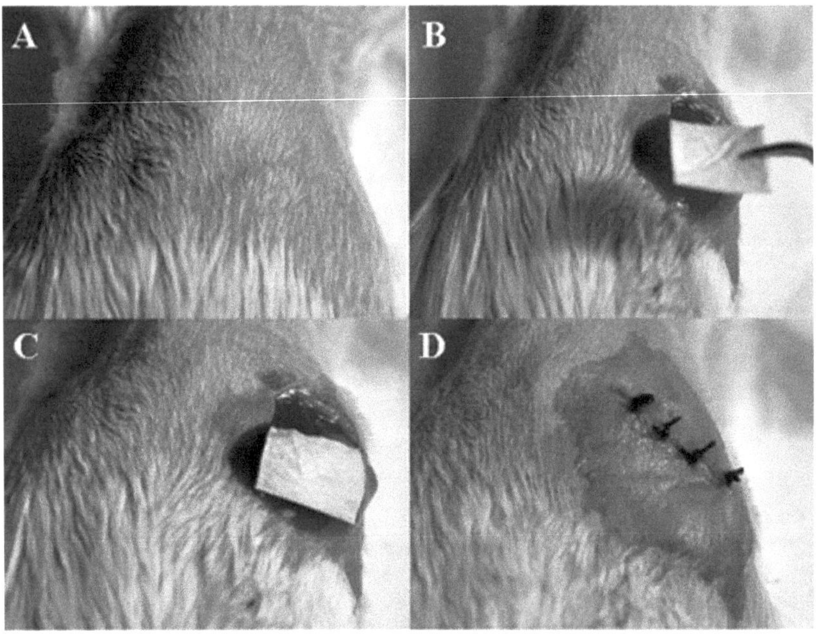

Fig. 2.3 Optical photograph showing the administration and therapeutic process of DOX-PLA nanofiber in mice: (A) anesthetized mouse, (B) exposure of the liver after laparotomy, (C) direct placement of the nanofiber on the tumor site, and (D) abdominal closure. *Reproduced with permission from S. Liu, G.Y. Zhou, D.X. Liu, Z.G. Xie, Y.B. Huang, X. Wang, W.B. Wu, X.B. Jing, Inhibition of orthotopic secondary hepatic carcinoma in mice by doxorubicin-loaded electrospun polylactide nanofibers. J. Mater. Chem. B 1 (2013) 101–109. Royal Society of Chemistry.*

Chemotherapy drugs shows extreme efficacy once it is administered in combination with other drugs commonly called combination chemotherapy. The logic behind combination chemotherapy is to use drugs that work by different mechanisms, thereby decreasing the probability of drug-resistant cells [49]. Ma et al. [50] have reported on the efficacy of cisplatin and curcumin (CUR)-loaded nanofibers for preventing cervical cancer recurrence. They have fabricated poly(ethylene oxide)/polylactide (PEO/PLA) nanofibers simultaneously loaded with cisplatin and CUR by electrospinning and exploited for combination therapy in vitro and in vivo. The results have revealed the supreme efficacy of the nanofiber composite in tumor inhibition compared to the control group and the same combination of drugs are administered through intravenous injection group.

Luo et al. [51] reported the antitumor efficacy of hydroxycamptothecin (HCPT)-loaded emulsion electrospun nanofibers in in vitro and in vivo trials. HCPT-loaded nanofibers exhibited 20 times enhanced anticancer efficacy against hepatocellular carcinoma cells (HepG2) compared to free HCPT during in vitro cytotoxicity tests. Same result was obtained in in vivo trials and the HCPT-loaded fibers exhibited a significantly higher tumor inhibition effect after intratumoral implantation compared with free HCPT, revealing the promising application of the system for local treatment of solid tumors.

Aggarwal et al. [52] studied the therapeutic efficacy of anticancer drug cisplatin-loaded poly-caprolactone chitosan composite nanofibers for local chemotherapy of cervical cancer in mice. The main benefit of nanofiber-based study is the site-specific superior availability of drug due to the mucoadhesive nature of nanofiber compared to the plain drug solution as well as the availability of cisplatin in its active state resulted in enhanced antitumor efficacy. Thus the study put forward a satisfactory approach for the targeted delivery of cancer therapeutics through vaginal routes for cervical cancer treatment.

In this book, we have focused on some of the recent studies describing the chemotherapeutic-loaded nanofibers as there are several reports depicting the in vitro anticancer efficacy of drug-loaded nanofibers [25, 53, 54]. Yohe et al [27] fabricated a drug-loaded three-dimensional (3D) superhydrophobic electrospun meshes to prevent locoregional recurrence in colorectal cancer. The dual functional 3D mesh will reinforce the surgical margin as well as locally elute drug for prolonged periods to prevent locoregional recurrence in colorectal cancer. Kumar et al. [55] evaluated the role of polymer cross-linking in drug release kinetics in a dual drug-loaded core-shell nanofibers with different extents of cross-linking and polymer composition as shown in Fig. 2.4. They have also studied the synergistic anticancer

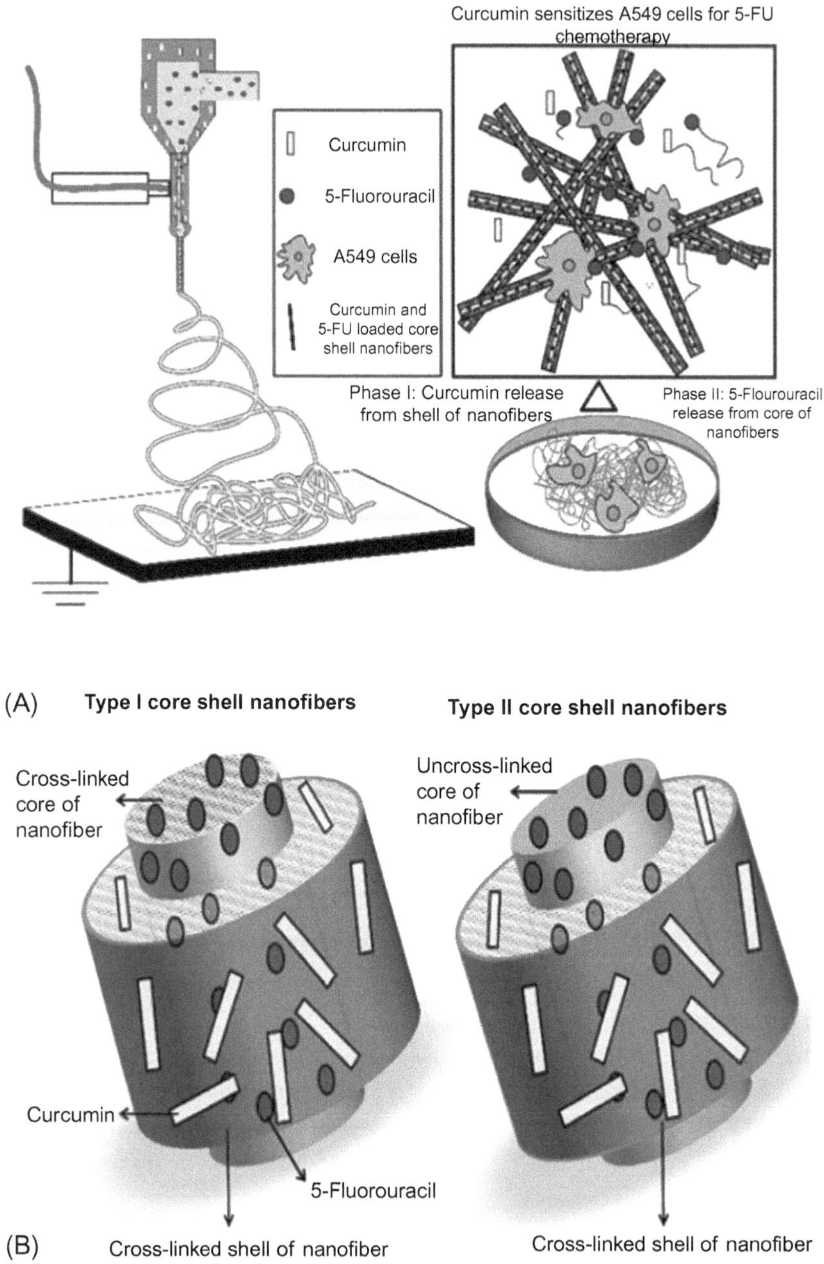

Fig. 2.4 Schematic representation showing (A) the fabrication of drug-loaded core-shell nanofibers by coaxial electrospinning and their in vitro anticancer efficacy, and (B) classification of nanofiber based on the change in polymer cross-linking. *Reproduced with permission from S.U. Kumar, I. Matai, P. Dubey, B. Bhushan, A. Sachdev, P. Gopinath, Differentially cross-linkable core-shell nanofibers for tunable delivery of anticancer drugs: synthesis, characterization and their anticancer efficacy. RSC Adv. 4 (2014) 38263–38272. Royal Society of Chemistry.*

efficacy of the two nanofiber systems and found that the two nanofibers exhibited contrasting drug release profile and resulted in different therapeutic efficacy at different time points against non-small cell lung cancer cells.

Qiu et al. [56] prepared DOX-loaded mesoporous silica nanoparticles (DOX@MSNs) encapsulated electrospun ploy(L-lactic acid) (PLLA) nanofibers for anticancer therapy. They have studied the in vitro DOX release profiles and anticancer efficacy and found that the DOX@MSNs-loaded nanofiber matrix exhibit sustained and prolonged drug release compared to the MSN-free counterparts. From the encouraging in vitro anticancer efficacy against Hela cells, they anticipate the potential application of the propped material for postsurgical breast cancer treatment. Ma et al. [57] reported a paclitaxel (PTX) loaded highly porous electrospun nanofibers for prostate cancer treatment. For that they have first fabricated the nanofiber using chitosan/polyethylene oxide (PEO) blend solution and then removed PEO with water. PTX was further loaded on to the porous nanofiber and the entire nanofiber matrix was encapsulated by hyaluronic acid by simple immersion technique to enhance the bioavailability. The in vitro anticancer results shows the efficacy against prostate cancer cells, which reveals the successfulness of using nanofiber matrix as a prospective material against prostate cancer.

2.3.2 Electrospun nanofibers for naturally derived chemotherapeutic delivery

Traditional anticancer drug formulations have their own shortcomings such as short half-life, unpredictable side effects, low therapeutic efficiency toward selected tumors, etc. Thus, numerous research have been conducted to develop natural material-based systems to improve the therapeutic efficacy against cancer. Among them green tea polyphenol (GTP) is one of the widely studied naturally derived materials for anticancer therapy. GTP accounts for about 10% of the dry weight of green tea and possess many vital functionalities such as active anti-oxidative potential, anti-inflammatory effect, and potential anticancer property. As a vital component of GTP, epigallocatechin-3-gallate (EGCG) has received a great deal of attention during the previous studies as an active anticancer agent. Moreover, the anticancer effect is successfully shown in various organs such as lung, esophagus, skin, liver, and stomach [58–60].

In a recent study, Shao et al. [61] have reported the anticancer effects of GTP encapsulated into biodegradable polymer nanofibers obtained by electrospinning process. They have developed polycaprolactone

(PCL)/multiwalled carbon nanotube (MWCNT) composite nanofibers with various concentration of GTP and found that the nanofiber system exhibited controlled release and extended stability of the encapsulated GTP. In vitro cytotoxicity evaluation of GTP showed that its inhibition effect to cancer cells was outstanding whereas the toxicity for normal cells was negligible with small amount of GTP as shown in Fig. 2.5. In addition to that the hydrophilicity of the nanofibers was also improved due to the addition of hydrophilic GTP. Hence the affinity between cells and nanofiber matrix were enhanced. The study indicated that the GTP-loaded polymer nanofibers with minimal side effects own a great prospective in the treatment of cancers. The exact mechanism of GTP-based cancer inhibition is not clear

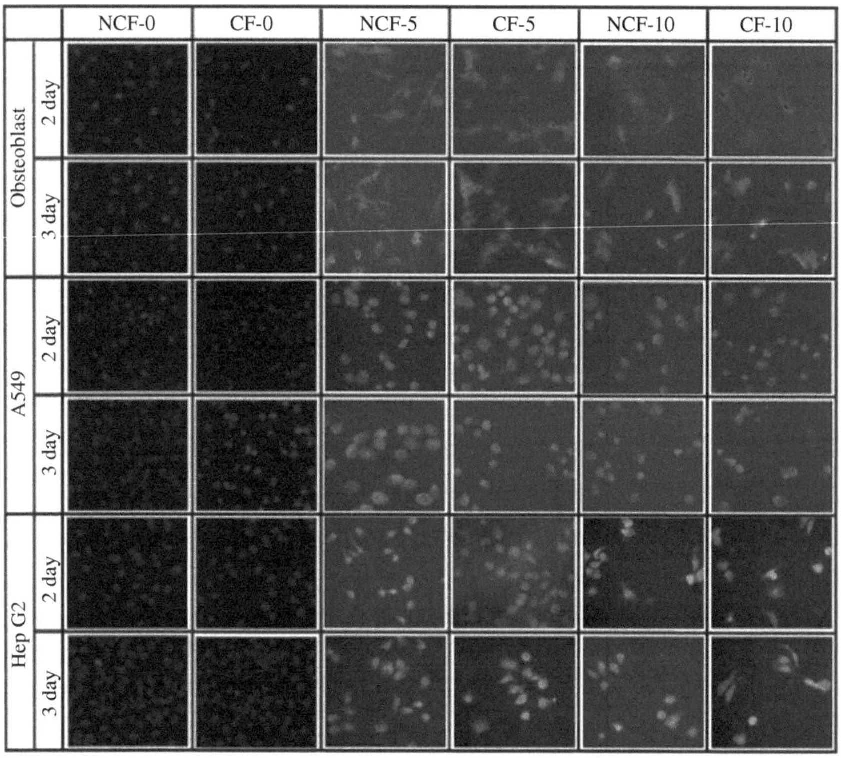

Fig. 2.5 Fluorescence microscope image of osteoblast, A549, and Hep G2 tumor cells cultured on PCL nanofiber meshes with GTP content 0% (NCF-0), 5% (NCF-5), 10% (NCF-10), and PCL/MWCNTs nanofiber meshes with GTP content 0% (CF-0), 5% (CF-0), 10% (CF-0) for 2 and 3 days. *Reproduced with permission from S.J. Shao, L. Li, G. Yang, J.R. Li, C. Luo, T. Gong, S.B. Zhou, Controlled green tea polyphenols release from electrospun PCL/MWCNTs composite nanofibers, Int. J. Pharm. 421 (2011) 310–320, Elsevier.*

till now. It has been put forward that the ability of GTP to subdue the formation of new blood vessels may be the possible reason for the inhibition of tumor growth and metastasis. However, the mechanism is very important and interesting and paves way for the future studies in cancer inhibition.

Similarly in another study, polymeric nanofiber system for the sustained release of piperine (PIP) for cancer therapy is introduced. PIP is an alkaloid derived from black pepper (Piper nigrum Linn) and long pepper (Piper longum Linn). It has been shown that PIP has antidepressant, antiarthritic, anti-inflammatory, antibacterial, and antioxidant effects. [62–64] It is also well known that phytochemicals or plant-derived nutritional agents like CUR and PIP cause fewer side effects as they are used as food ingredients so the human body is already adapted to them. Jain et al. [65] have reported PIP encapsulated PCL-GEL nanofibers for localized controlled delivery of PIP to the tumor site for use as postsurgical implants. The PCL/GEL nanofibers were loaded with up to 30 wt% PIP, and studied the *in vitro* release kinetics. The nanofiber system exhibited a sustained release of the PIP with 50% release in 3 days from the PCL/GEL (50:50 by weight) blend fibers. The reduced viability and growth of HeLa and MCF-7 cancer cells on the PIP eluting nanofibers demonstrated anticancer activity in vitro. The proliferation of noncancerous cells such as NIH3T3 cells and human mesenchymal stem cells was affected to a lesser extent. Flow cytometry revealed that the released PIP induced the generation of reactive oxygen species (ROS) and cell cycle arrest in the G2/M phase, leading to the death of cancer cells. The findings of this study suggest that PIP-loaded nanofiber mats could be developed into implantable biodegradable patches for use in postsurgical cancer treatment.

CUR, which is the main phenolic pigment extracted from turmeric, has shown extensive biological benefits. Widespread researches on the biological activity of CUR have indicated its antibacterial, anticancer, anti-inflammation, and antioxidant activities [66–68]. Antitumor properties of CUR have been advantageous to several types of cancer such as stomach, colon, liver, fibroblast, and breast cancer. CUR is a nontoxic multi-target natural compound which can profoundly prevent drug resistance and also sensitize cancer cells to chemotherapeutic agents. Sedghi et al. [69] reported on the application of biomedicated nanofibers for preventing wound infection and local chemotherapy. For that first they prepared functionalized graphene oxide silica curcumin nanocomposite (fGO-Si-CUR) and electrospun with PVA and chitosan CS to exploit its antibacterial and anticancer properties. In vitro study result confirmed that the GO functionalized curcumin nanocomposite nanofiber exhibit a sustained drug release profile compared to their unmodified counterpart CUR-PVA-CS nanofiber. In vitro anticancer studies on MCF 7, HEPG2, and L929 cell lines have shown the enhanced anticancer ability and have concluded that fGO-Si-CUR-loaded nanofibers can be used as a potential material for postoperative chemotherapy.

Plant polyphenols are gaining wide attention due to their unique neuroprotective, cardioprotective, and chemopreventive properties [70]. They are referred to as diverse group of compounds containing several phenolic functionalities and in particular they interfere in several steps of tumor inhibition [71]. Kim et al. have reported on the potential of plant polyphenol-loaded PCL nanofibers for prolonged anticancer drug delivery after the surgery. They have used (−)-epigallocatechin-3-*O*-gallate (EGCG) and caffeic acid (CA) as plant polyphenols to make the composite fibers and found that they exhibit a controlled release of EGCG and CA from the nanofiber matrix without any initial burst release. The in vitro anticancer study also showed a greater cytotoxicity due to the apoptosis induced by the H_2O_2 produced by EGCG and CA. Thus the study reveals a promising platform for anticancer therapy.

2.3.3 Electrospun nanofibers thermo-chemotherapy

In recent years, combination therapies such as chemotherapy with radiation therapy, radiation therapy with thermo therapy, etc., have proven to be extremely important for the eradication of several types of tumors. Here we are mainly focusing on nanofibers for hyperthermic chemotherapy application. Magnetic nanoparticles have been reported as the key material for magnetic hyperthermia application, in which body tissues are exposed to slightly higher temperatures in order to damage or kill cancer cells by inducing apoptosis, or to make cancer cells more sensitive to the effects of radiation and/or certain anticancer drugs [72]. Fig. 2.6 indicates the advantage of combined application of hyperthermia and chemotherapy for the synergistic anticancer effects.

Kim et al. [41] have developed temperature-responsive polymer nanofibers incorporating anticancer drug DOX and magnetic nanoparticle to enable hyperthermic chemotherapeutic application. The lower critical solution temperature of the thermosensitive polymer was adjusted to hyperthermic temperature so as to enable a switchable DOX release. The smart nanofiber system exhibited a synergistic anticancer efficacy against human

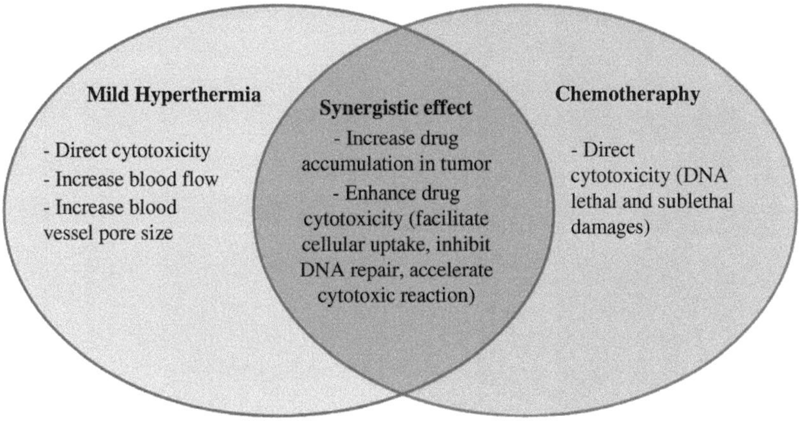

Fig. 2.6 Mechanisms responsible for the synergistic effect of the combined therapy. *Reproduced from A. Hervault, N.T.K. Thanh, Magnetic nanoparticle-based therapeutic agents for thermo-chemotherapy treatment of cancer. Nanoscale 6 (2014) 11553–11573. Royal Society of Chemistry.*

melanoma cell lines. GhavamiNejad et al. [23] developed a catecholic nanofiber which can bind iron oxide nanoparticle (IONP) as well as anticancer drug Bortezomib (BTZ) through a catechol metal binding mechanism. The drug binding was made in such a way that it will be released in a cancer environment (acidic pH). They studied the anticancer efficacy of the smart nanofibers against different cancer cell lines and found the synergistic efficacy of the nanosystem. Thus they have concluded the nanofiber matrix as a prospective material for combined application of hyperthermia and chemotherapy with hyperthermic treatment time being <5 min (Fig. 2.7).

Sasikala et al. [24] have reported an implantable smart magnetic nanofiber device for endoscopic hyperthermia treatment and tumor-triggered controlled drug release. This device is synthesized using a two-component smart nanofiber matrix from IONPs as well as a chemotherapeutic drug BTZ. IONP-incorporated nanofiber matrix was developed by electrospinning the biocompatible and bioresorbable polymer poly(D,L-lactide-co-glycolide) (PLGA) and the tumor-triggered anticancer drug delivery is realized by exploiting the mussel inspired surface functionalization to conjugate the borate-containing anticancer drug BTZ through a catechol metal binding in a pH-sensitive manner. Thus the obtained implantable smart magnetic nanofiber devices were exploited for both hyperthermia application in an alternating magnetic field (AMF) and cancer cell-specific controlled drug release for synergistic cancer therapy. The results confirm that the

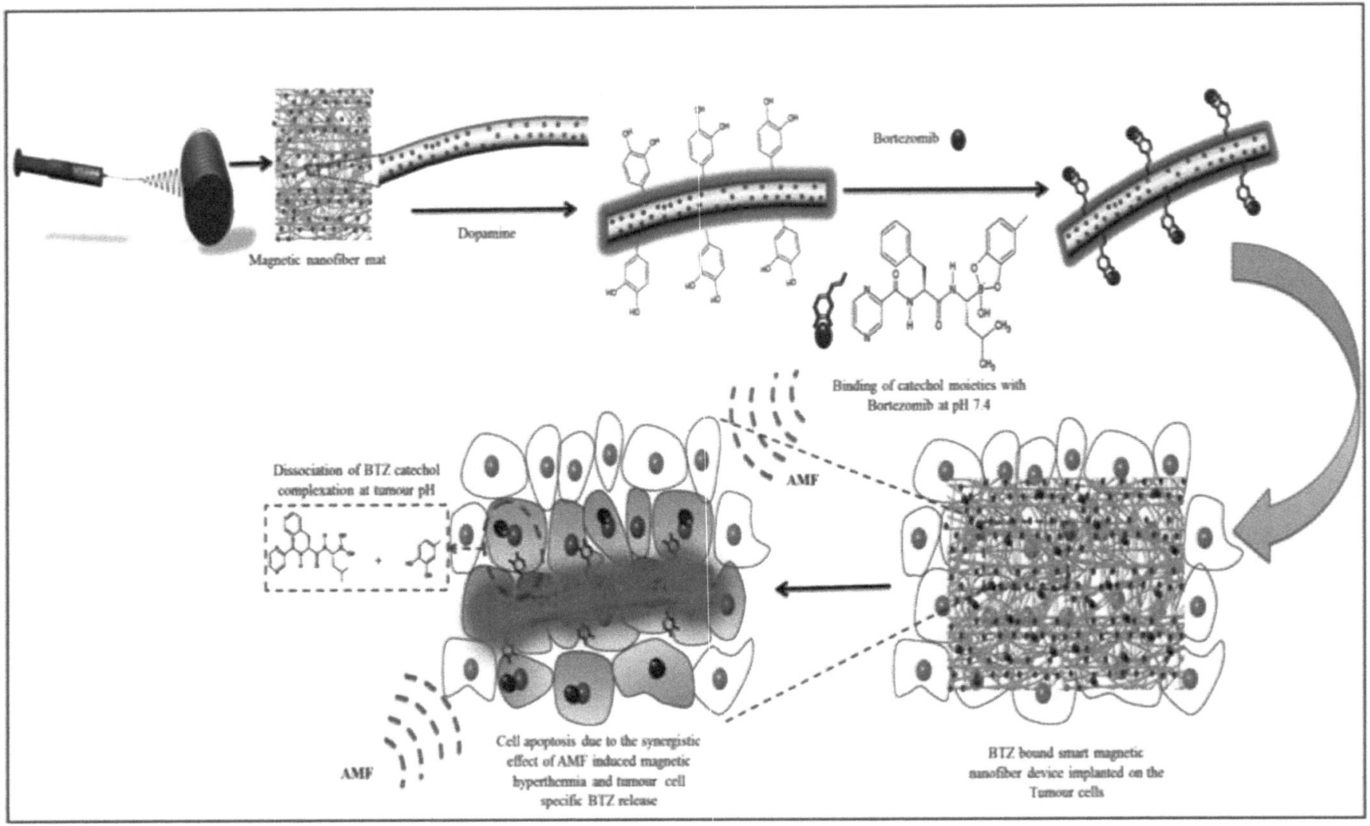

Fig. 2.7 Scheme illustration showing smart nanoplatform responsive to a magnetic field to administer both hyperthermia and pH-dependent anticancer drug release for the synergistic anticancer treatment. *Reproduced with permission from A. GhavamiNejad, A.R.K. Sasikala, A.R. Unnithan, R.G. Thomas, Y.Y. Jeong, M. Vatankhah-Varnoosfaderani, F.J. Stadler, C.H. Park, C.S. Kim, Mussel-inspired electrospun smart magnetic nanofibers for hyperthermic chemotherapy, Adv. Funct. Mater. 25 (2015) 2867–2875, Elsevier.*

BTZ-loaded mussel-inspired magnetic nanofiber matrix (BTZ-MMNF) is very beneficial not only due to their higher therapeutic efficacy and low toxicity toward normal cells, but also due to the availability of magnetic nanoparticle for repeated hyperthermia application and tumor-triggered controlled drug release.

Photothermal therapy is also a promising alternative to traditional cancer therapies; which uses a laser-induced hyperthermia ablation [73, 74]. Zhang et al. [75] reported a single system using MWCNTs and anticancer drug DOX-loaded PLLA nanofibers for combined photothermal and chemotherapy application as shown in Fig. 2.8. According to their study the nanofiber displayed enhanced anticancer efficacy in both in vitro and in vivo application due to the synergistic effect of DOX and photothermal therapy. This is due to the NIR (near-infrared)-responsive controlled DOX release from the multifunctional nanofiber matrix with maximum drug concentration in tumor region as it has been used as a local delivery system along with the availability of MWCNTs in the tumor surroundings for repeated localized heating. Moreover, the increased cell binding due to the highly porous nanofiber also resulted in enhanced anticancer efficacy. Thus the multifunctional nanofiber matrix for thermo-chemotherapy application results is very promising for easily accessible tumors and highly aggressive tumors. More research in this field will benefit the future of cancer nanomedicine.

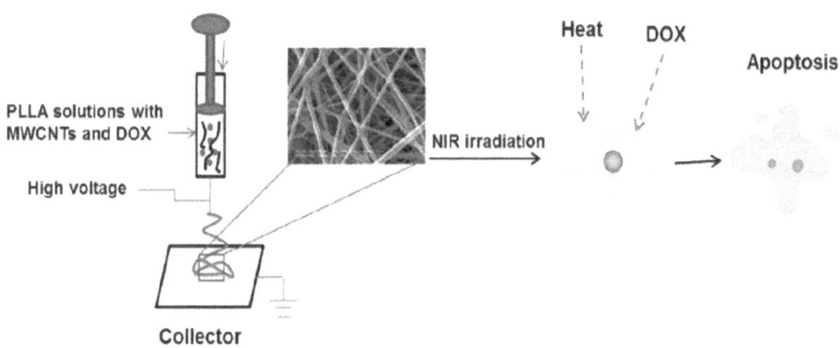

Fig. 2.8 Scheme showing multiwalled carbon nanotubes (MWCNTs) and anticancer drug DOX-loaded PLLA nanofibers for combined photothermal and chemotherapy application. *Reproduced with permission from Z. Zhang, S. Liu, H. Xiong, X. Jing, Z. Xie, X. Chen, Y. Huang, Electrospun PLA/MWCNTs composite nanofibers for combined chemo- and photothermal therapy. Acta Biomater. 26 (2015) 115–123. Elsevier.*

2.3.4 Electrospun nanofibers for therapy and lost tissue regeneration

Recently, nanofiber-based cancer therapeutics are receiving great attention for localized cancer treatment. Some reports have come up with a great idea of cancer therapy and lost tissue regeneration in a single platform using versatile nanofiber-based formulations. Wang et al. [76] have reported multifunctional composite nanofiber for skin cancer therapy and tissue regeneration to heal the tumor-induced cutaneous wounds. They have used a modified electrospinning method to incorporate Cu_2S nanoflowers into PLA/PCL biopolymer to obtain CS-PLA/PCL micropatterned nanomembrane. The uniformly distributed Cu_2S nanoparticles offer excellent photothermal ability under NIR radiation and the ECM-like structure of the nanofiber support the enhanced cell adhesion, proliferation, and migration of skin cells. The in vivo study result displayed the effective tumor inhibition and full thickness skin wound healing property. Thus the study offers reliable strategy for localized tumor therapy and lost tissue regeneration.

Another study by Sasikala et al. [22] have reported a multifaceted implantable anticancer device (IAD) using nanofiber matrix for postsurgical breast cancer treatment as shown in Fig. 2.9. They have first fabricated the superparamagnetic iron oxide graphene oxide doxorubicin nanocomposite and electrospun the nanocomposite with polyurethane (PU) to obtain the IAD. The IAD were exploited for sustained and tumor-specific anticancer drug release and repeated hyperthermic application in an AMF against murine breast cancer cell lines. The IAD also exhibited a noninvasive monitoring capability through magnetic resonance imaging. The IAD was further exploited to evaluate the in vitro adipogenesis. Thus the study put forward a new concept for overcoming the breast cancer recurrence and completely cures the lost/dead cells by adipogenesis.

2.4 Conclusion and challenges

Electrospun nanofibers have shown great potential for the local delivery of cancer therapeutic agents by overcoming the obstacles faced by systemic delivery and result in a more successful outcome. The advancement in local drug delivery using nanofiber may permit the recommendation of drugs that might have previously failed with a systemic treatment. Moreover, several kinds of drugs/or drug-loaded composite nanosystems could be easily

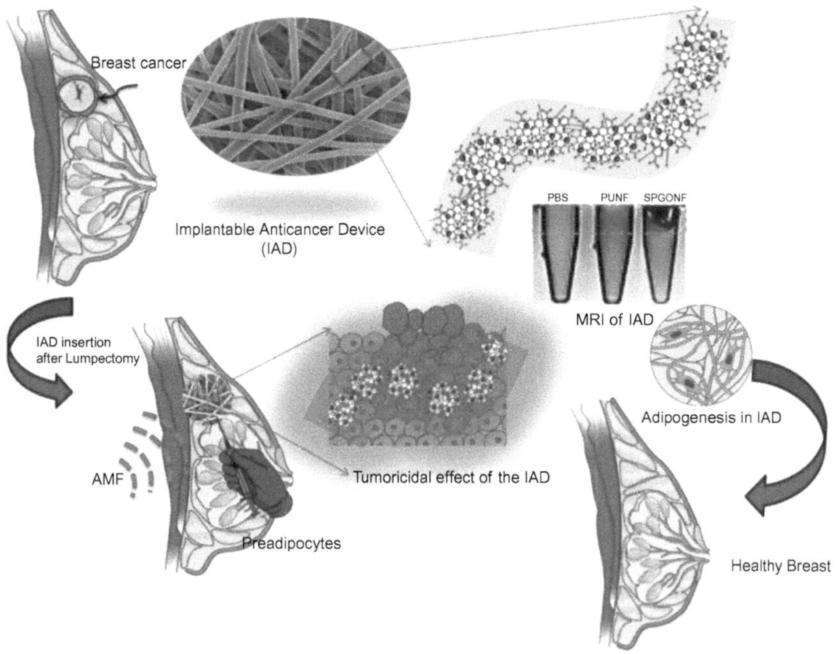

Fig. 2.9 Schematic representation showing the overall idea of IAD for breast cancer theranostics and breast tissue reconstruction. *Reproduced with permission from A.R.K. Sasikala, A.R. Unnithan, R.G. Thomas, S.W. Ko, Y.Y. Jeong, C.H. Park and C.S. Kim, Multifaceted implantable anticancer device for potential postsurgical breast cancer treatment: a single platform for synergistic inhibition of local regional breast cancer recurrence, surveillance, and healthy breast reconstruction, Adv. Funct. Mater. 28 (2018), WILEY-VCH Verlag GmbH & Co. KGaA, Weinheim.*

incorporated into a nanofiber system by direct as well as indirect method of fabrication and the nanofiber would protect the drug from corrosion due to gastric acid or enzymes in the body as well as maintain the bioavailability. The nanofiber-based implantable delivery system can be easily exploited for combination therapy by properly incorporating the therapeutic cargo into the nanofiber system. Recent development has shown that the nanofiber-based system can be used as a platform to totally eradicate the disease and to rejuvenate the affected/lost tissues for the complete recovery. Thus electrospun nanofiber-based implantable systems offer versatile and tailorable approach to fight cancer and the scope of nanofibers in cancer nanomedicine is immense.

References

[1] H.D. Wang, M. Naghavi, C. Allen, R.M. Barber, Z.A. Bhutta, A. Carter, D.C. Casey, F.J. Charlson, A.Z. Chen, M.M. Coates, M. Coggeshall, L. Dandona, D.J. Dicker, H.E. Erskine, A.J. Ferrari, C. Fitzmaurice, K. Foreman, M.H. Forouzanfar, M. S. Fraser, N. Pullman, P.W. Gething, E.M. Goldberg, N. Graetz, J.A. Haagsma, S.I. Hay, C. Huynh, C. Johnson, N.J. Kassebaum, Y. Kinfu, X.R. Kulikoff, M. Kutz, H.H. Kyu, H.J. Larson, J. Leung, X.F. Liang, S.S. Lim, M. Lind, R. Lozano, N. Marquez, G.A. Mensah, J. Mikesell, A.H. Mokdad, M.D. Mooney, G. Nguyen, E. Nsoesie, D.M. Pigott, C. Pinho, G.A. Roth, J.A. Salomon, L. Sandar, N. Silpakit, A. Sligar, R.J.D. Sorensen, J. Stanaway, C. Steiner, S. Teeple, B.A. Thomas, C. Troeger, A. VanderZanden, S.E. Vollset, V. Wanga, H. A. Whiteford, T. Wolock, L. Zoeckler, K.H. Abate, C. Abbafati, K.M. Abbas, F. Abd-Allah, S.F. Abera, D.M.X. Abreu, L.J. Abu-Raddad, G.Y. Abyu, T. Achoki, A.L. Adelekan, Z. Ademi, A.K. Adou, J.C. Adsuar, K.A. Afanvi, A. Afshin, E. E. Agardh, A. Agarwal, A. Agrawal, A.A. Kiadaliri, O.N. Ajala, A.S. Akanda, R. O. Akinyemi, T.F. Akinyemiju, N. Akseer, F.H. Al Lami, S. Alabed, Z. Al-Aly, K. Alam, N.K.M. Alam, D. Alasfoor, S.F. Aldhahri, R.W. Aldridge, M.A. Alegretti, A.V. Aleman, Z.A. Alemu, L.T. Alexander, S. Alhabib, R. Ali, A. Alkerwi, F. Alla, P. Allebeck, R. Al-Raddadi, U. Alsharif, K.A. Altirkawi, E.A. Martin, N. Alvis-Guzman, A.T. Amare, A.K. Amegah, E.A. Ameh, H. Amini, W. Ammar, S. M. Amrock, H.H. Andersen, B. Anderson, G.M. Anderson, C.A.T. Antonio, A. F. Aregay, J. Arnlov, V.S.A. Arsenijevic, A. Artaman, H. Asayesh, R.J. Asghar, S. Atique, E.F.G.A. Avokpaho, A. Awasthi, P. Azzopardi, U. Bacha, A. Badawi, M. C. Bahit, K. Balakrishnan, A. Banerjee, A. Barac, S.L. Barker-Collo, T. Barnighausen, L. Barregard, L.H. Barrero, A. Basu, S. Basu, Y.T. Bayou, S. Bazargan-Hejazi, J. Beardsley, N. Bedi, E. Beghi, H.A. Belay, B. Bell, M.L. Bell, A.K. Bello, D.A. Bennett, I.M. Bensenor, A. Berhane, E. Bernabe, B.D. Betsu, A. S. Beyene, N. Bhala, A. Bhalla, S. Biadgilign, B. Bikbov, A.A. Bin Abdulhak, B. J. Biroscak, S. Biryukov, E. Bjertness, J.D. Blore, C.D. Blosser, M.A. Bohensky, R. Borschmann, D. Bose, R.R.A. Bourne, M. Brainin, C.E.G. Brayne, A. Brazinova, N.J.K. Breitborde, H. Brenner, J.D. Brewer, A. Brown, J. Brown, T. S. Brugha, G.C. Buckle, Z.A. Butt, B. Calabria, I.R. Campos-Novato, J. C. Campuzano, J.R. Carapetis, R. Cardenas, D. Carpenter, J.J. Carrero, C. A. Castaneda-Oquela, J.C. Rivas, F. Catala-Lopez, F. Cavalleri, K. Cercy, J. Cerda, W.Q. Chen, A. Chew, P.P.C. Chiang, M. Chibalabala, C.E. Chibueze, O. Chimed-Ochir, V.H. Chisumpa, J.Y.J. Choi, R. Chowdhury, H. Christensen, D.J. Christopher, L.G. Ciobanu, M. Cirillo, A.J. Cohen, V. Colistro, M. Colomar, S.M. Colquhoun, C. Cooper, L.T. Cooper, M. Cortinovis, B.C. Cowie, J. A. Crump, J. Damsere-Derry, H. Danawi, R. Dandona, F. Daoud, S.C. Darby, P. I. Dargan, J. das Neves, G. Davey, A.C. Davis, D.V. Davitoiu, E.F. de Castro, P. de Jager, D. De Leo, L. Degenhardt, R.P. Dellavalle, K. Deribe, A. Deribew, S. D. Dharmaratne, P.K. Dhillon, C. Diaz-Torne, E.L. Ding, K.P.B. dos Santos, E. Dossou, T.R. Driscoll, L.L. Duan, M. Dubey, B. Bartholow, R.G. Ellenbogen, C. Lycke, I. Elyazar, A.Y. Endries, S.P. Ermakov, B. Eshrati, A. Esteghamati, K. Estep, I.D.A. Faghmous, S. Fahimi, E. Jose, T.A. Farid, C.S.E.S. Farinha, A. Faro, M.S. Farvid, F. Farzadfar, V.L. Feigin, S.M. Fereshtehnejad, J. G. Fernandes, J.C. Fernandes, F. Fischer, J.R.A. Fitchett, A. Flaxman, N. Foigt, F. G.R. Fowkes, E.B. Franca, R.C. Franklin, J. Friedman, J. Frostad, T. Hirst, N. D. Futran, S.L. Gall, K. Gambashidze, A. Gamkrelidze, P. Ganguly, F.G. Gankpe, T. Gebre, T.T. Gebrehiwot, A.T. Gebremedhin, A.A. Gebru, J.M. Geleijnse, B. D. Gessner, A.G. Ghoshal, K.B. Gibney, R.F. Gillum, S. Gilmour, A.Z. Giref,

M. Giroud, M.D. Gishu, G. Giussani, E. Glaser, W.W. Godwin, H. Gomez-Dantes, P. Gona, A. Goodridge, S.V. Gopalani, R.A. Gosselin, C.C. Gotay, A. Goto, H. N. Gouda, F. Greaves, H.C. Gugnani, R. Gupta, R. Gupta, V. Gupta, R. A. Gutierrez, N. Hafezi-Nejad, D. Haile, A.D. Hailu, G.B. Hailu, Y.A. Halasa, R. R. Hamadeh, S. Hamidi, J. Hancock, A.J. Handal, G.J. Hankey, Y.T. Hao, H. L. Harb, S. Harikrishnan, J.M. Haro, R. Havmoeller, S.R. Heckbert, I.B. Heredia-Pi, P. Heydarpour, H.B.M. Hilderink, H.W. Hoek, R.S. Hogg, M. Horino, N. Horita, H.D. Hosgood, P.J. Hotez, D.G. Hoy, M. Hsairi, A.S. Htet, M.M. T. Htike, G.Q. Hu, C. Huang, H. Huang, L. Huiart, A. Husseini, I. Huybrechts, G. Huynh, K.M. Iburg, K. Innos, M. Inoue, V.J. Iyer, T.A. Jacobs, K.H. Jacobsen, N. Jahanmehr, M.B. Jakovljevic, P. James, M. Javanbakht, S.P. Jayaraman, A. U. Jayatilleke, P. Jeemon, P.N. Jensen, V. Jha, G. Jiang, Y. Jiang, T. Jibat, A. Jimenez-Corona, J.B. Jonas, T.K. Joshi, Z. Kabir, R. Karnak, H.D. Kan, S. Kant, A. Karch, C.K. Karema, C. Karimkhani, D. Karletsos, G. Karthikeyan, A. Kasaeian, M. Katibeh, A. Kaul, N. Kawakami, J.F. Kayibanda, P.N. Keiyoro, L. Kemmer, A. H. Kemp, A.P. Kengne, A. Keren, M. Kereselidze, C.N. Kesavachandran, Y. S. Khader, I.A. Khalil, A.R. Khan, E.A. Khan, Y.H. Khang, S. Khera, T.A. M. Khoja, C. Kieling, D. Kim, Y.J. Kim, B.M. Kissela, N. Kissoon, L.D. Knibbs, A.K. Knudsen, Y. Kokubo, D. Kolte, J.A. Kopec, S. Kosen, P.A. Koul, A. Koyanagi, N.H. Krog, B.K. Defo, B.K. Bicer, A.A. Kudom, E.J. Kuipers, V. S. Kulkarni, G.A. Kumar, G.F. Kwan, A. Lal, D.K. Lal, R. Lalloo, H. Lam, J. O. Lam, S.M. Langan, V.C. Lansingh, A. Larsson, D.O. Laryea, A.A. Latif, A.E. B. Lawrynowicz, J. Leigh, M. Levi, Y. Li, M.P. Lindsay, S.E. Lipshultz, P.Y. Liu, S. Liu, Y. Liu, L.T. Lo, G. Logroscino, P.A. Lotufo, R.M. Lucas, R. Lunevicius, R.A. Lyons, S. Ma, V.M.P. Machado, M.T. Mackay, J.H. MacLachlan, H.M.A. El Razek, M.M.A. El Razek, M. Majdan, A. Majeed, R. Malekzadeh, W.A. A. Manamo, J. Mandisarisa, S. Mangalam, C.C. Mapoma, W. Marcenes, D. J. Margolis, G.R. Martin, J. Martinez-Raga, M.B. Marzan, F. Masiye, A.J.M. Jones, J. Massano, R. Matzopoulos, B.M. Mayosi, S.T. McGarvey, J.J. McGrath, M. Mckee, B.J. McMahon, P.A. Meaney, A. Mehari, M.M. Mehndiratta, F. Mena-Rodriguez, A.B. Mekonnen, Y.A. Melaku, P. Memiah, Z.A. Memish, W. Mendoza, A. Meretoja, T.J. Meretoja, F.A. Mhimbira, R. Micha, T.R. Miller, M. Mirarefin, A. Misganaw, C.N. Mock, K.A. Mohammad, A. Mohammadi, S. Mohammed, V. Mohan, G.L.D. Mola, L. Monasta, J.C.M. Hernandez, P. Montero, M. Montico, T.J. Montine, M. Moradi-Lakeh, L. Morawska, K. Morgan, R. Mori, D. Mozaffarian, U. Mueller, G.V.S. Murthy, S. Murthy, K. I. Musa, J.B. Nachega, G. Nagel, K.S. Naidoo, N. Naik, L. Naldi, V. Nangia, D. Nash, C. Nejjari, S. Neupane, C.R. Newton, J.N. Newton, M. Ng, F. N. Ngalesoni, J.D. Ngirabega, Q. Le Nguyen, M.I. Nisar, P.M.N. Pete, M. Nomura, O.F. Norheim, P.E. Norman, B. Norrving, L. Nyakarahuka, F. A. Ogbo, T. Ohkubo, F.A. Ojelabi, P.R. Olivares, B.O. Olusanya, J.O. Olusanya, J.N. Opio, E. Oren, A. Ortiz, M. Osman, E. Ota, R. Ozdemir, M. Pa, J. D. Pandian, P.R. Pant, C. Papachristou, E.K. Park, J.H. Park, C.D. Parry, M. Parsaeian, A.J.P. Caicedo, S.B. Patten, G.C. Patton, V.K. Paul, N. Pearce, J. M. Pedro, L.P. Stokic, D.M. Pereira, N. Perico, K. Pesudovs, M. Petzold, M. R. Phillips, F.B. Piel, J.D. Pillay, D. Plass, J.A. Platts-Mills, S. Polinder, C.A. Pope, S. Popova, R.G. Poulton, F. Pourmalek, D. Prabhakaran, M. Qorbani, J. Quame-Amaglo, D.A. Quistberg, A. Rafay, K. Rahimi, V. Rahimi-Movaghar, M. Rahman, M.H.U. Rahman, S.U. Rahman, R.K. Rai, Z. Rajavi, S. Rajsic, M. Raju, I. Rakovac, S.M. Rana, C.L. Ranabhat, T. Rangaswamy, P. Rao, S.R. Rao, A. H. Refaat, J. Rehm, M.B. Reitsma, G. Remuzzi, S. Resnikofff, A.L. Ribeiro, S. Ricci, M.J.R. Blancas, B. Roberts, A. Roca, D. Rojas-Rueda, L. Ronfani,

G. Roshandel, D. Rothenbacher, A. Roy, N.K. Roy, G.M. Ruhago, R. Sagar, S. Saha, R. Sahathevan, M.M. Saleh, J.R. Sanabria, M.D. Sanchez-Nino, L. Sanchez-Riera, I. S. Santos, R. Sarmiento-Suarez, B. Sartorius, M. Satpathy, M. Savic, M. Sawhney, M. P. Schaub, M.I. Schmidt, I.J.C. Schneider, B. Schottker, A.E. Schutte, D.C. Schwebel, S. Seedat, S.G. Sepanlou, E.E. Servan-Mori, K.A. Shackelford, G. Shaddick, A. Shaheen, S. Shahraz, M.A. Shaikh, M. Shakh-Nazarova, R. Sharma, J. She, S. Sheikhbahaei, J.B. Shen, Z.Y. Shen, D.S. Shepard, K.N. Sheth, B.P. Shetty, P. L. Shi, K. Shibuya, M.J. Shin, R. Shiri, I. Shiue, M.G. Shrime, I.D. Sigfusdottir, D. H. Silberberg, D.A.S. Silva, D.G.A. Silveira, J.I. Silverberg, E.P. Simard, A. Singh, G.M. Singh, J.A. Singh, O.P. Singh, P.K. Singh, V. Singh, S. Soneji, K. Soreide, J. B. Soriano, L.A. Sposato, C.T. Sreeramareddy, V. Stathopoulou, D.J. Stein, M. B. Stein, S. Stranges, K. Stroumpoulis, B.F. Sunguya, P. Sur, S. Swaminathan, B. L. Sykes, C.E.I. Szoeke, R. Tabares-Seisdedos, K.M. Tabb, K. Takahashi, J. S. Takala, R.T. Talongwa, N. Tandon, M. Tavakkoli, B. Taye, H.R. Taylor, B.J. T. Ao, B.A. Tedla, W.M. Tefera, M. Ten Have, A.S. Terkawi, F.H. Tesfay, G. A. Tessema, A.J. Thomson, A.L. Thorne-Lyman, A.G. Thrift, G.D. Thurston, T. Tillmann, D.L. Tirschwell, M. Tonelli, R. Topor-Madry, F. Topouzis, J.A. T. Nx, J. Traebert, B.X. Tran, T. Truelsen, U. Trujillo, A.K. Tura, E.M. Tuzcu, U.S. Uchendu, K.N. Ukwaja, E.A. Undurraga, O.A. Uthman, R. Van Dingenen, A. Van Donkelaar, T. Vasankari, A.M.N. Vasconcelos, N. Venketasubramanian, R. Vidavalur, L. Vijayakumar, S. Villalpando, F.S. Violante, V.V. Vlassov, J. A. Wagner, G.R. Wagner, M.T. Wallin, L.H. Wang, D.A. Watkins, S. Weichenthal, E. Weiderpass, R.G. Weintraub, A. Werdecker, R. Westerman, R. A. White, T. Wijeratne, J.D. Wilkinson, H.C. Williams, C.S. Wiysonge, S. M. Woldeyohannes, C.D.A. Wolfe, S.H. Won, J.Q. Wong, A.D. Woolf, D. Xavier, Q.Y. Xiao, G.L. Xu, B. Yakob, A.Z. Yalew, L.L. Yan, Y.C. Yano, M. Yaseri, P. Ye, H.G. Yebyo, P. Yip, B.D. Yirsaw, N. Yonemoto, G. Yonga, M. Z. Younis, S.C. Yu, Z. Zaidi, M.E.S. Zaki, F. Zannad, D.E. Zavala, H. Zeeb, B. M. Zeleke, H. Zhang, S. Zodpey, D. Zonies, L.J. Zuhlke, T. Vos, A.D. Lopez, C.J. L. Murray, G.M.A.C. Death, Global, regional, and national life expectancy, all-cause mortality, and cause-specific mortality for 249 causes of death, 1980-2015: a systematic analysis for the Global Burden of Disease Study 2015, Lancet 388 (2016) 1459–1544.

[2] M.A. Zeeshan, K.Y. Shou, K.M. Sivaraman, T. Wuhrmann, S. Pane, E. Pellicer, B. J. Nelson, Nanorobotic drug delivery if I only had a heart, Mater. Today 14 (2011) 54.

[3] R. Tong, D.S. Kohane, New strategies in cancer nanomedicine, Annu. Rev. Pharmacol. 56 (2016) 41–57.

[4] C. von Roemeling, W. Jian, C.K. Chan, I.L. Weissman, B.Y.S. Kim, Breaking down the barriers to precision cancer nanomedicine, Trends Biotechnol. 35 (2017) 159–171.

[5] J.J. Shi, P.W. Kantoff, R. Wooster, O.C. Farokhzad, Cancer nanomedicine: progress, challenges and opportunities, Nat. Rev. Cancer 17 (2017) 20–37.

[6] B. Pelaz, C.H. Alexiou, R.A. Alvarez-Puebla, F. Alves, A.M. Andrews, S. Ashraf, L. P. Balogh, L. Ballerini, A. Bestetti, C. Brendel, S. Bosi, M. Carril, W.C.W. Chan, C. Y. Chen, X.D. Chen, X.Y. Chen, Z. Cheng, D.X. Cui, J.Z. Du, C. Dullin, A. Escudero, N. Feliu, M.Y. Gao, M. George, Y. Gogotsi, A. Grunweller, Z. W. Gu, N.J. Halas, N. Hampp, R.K. Hartmann, M.C. Hersam, P. Hunziker, J. Jian, X.Y. Jiang, P. Jungebluth, P. Kadhiresan, K. Kataoka, A. Khademhosseini, J. Kopecek, N.A. Kotov, H.F. Krug, D.S. Lee, C.M. Lehr, K.W. Leong, X.J. Liang, M.L. Lim, L.M. Liz-Marzan, X.M. Ma, P. Macchiarini, H. Meng, H. Mohwald, P. Mulvaney, A.E. Nel, S.M. Nie, P. Nordlander, T. Okano, J. Oliveira, T. H. Park, R.M. Penner, M. Prato, V. Puntes, V.M. Rotello, A. Samarakoon, R. E. Schaak, Y.Q. Shen, S. Sjoqvist, A.G. Skirtach, M.G. Soliman, M.M. Stevens, H. W. Sung, B.Z. Tang, R. Tietze, B.N. Udugama, J.S. VanEpps, T. Weil, P.S. Weiss,

I. Willner, Y.Z. Wu, L.L. Yang, Z. Yue, Q. Zhang, Q. Zhang, X.E. Zhang, Y.L. Zhao, X. Zhou, W.J. Parak, Diverse applications of nanomedicine, ACS Nano 11 (2017) 2313–2381.
[7] S. Tran, P.J. DeGiovanni, B. Piel, P. Rai, Cancer nanomedicine: a review of recent success in drug delivery, Clin. Transl. Med. 6 (2017).
[8] Y. Malam, M. Loizidou, A.M. Seifalian, Liposomes and nanoparticles: nanosized vehicles for drug delivery in cancer, Trends Pharmacol. Sci. 30 (2009) 592–599.
[9] S.Y. Madani, N. Naderi, O. Dissanayake, A. Tan, A.M. Seifalian, A new era of cancer treatment: carbon nanotubes as drug delivery tools, Int. J. Nanomedicine 6 (2011) 2963–2979.
[10] T.M. Allen, P.R. Cullis, Drug delivery systems: entering the mainstream, Science 303 (2004) 1818–1822.
[11] J. Nicolas, S. Mura, D. Brambilla, N. Mackiewicz, P. Couvreur, Design, functionalization strategies and biomedical applications of targeted biodegradable/biocompatible polymer-based nanocarriers for drug delivery, Chem. Soc. Rev. 42 (2013) 1147–1235.
[12] K. Kataoka, G.S. Kwon, M. Yokoyama, T. Okano, Y. Sakurai, Block-copolymer micelles as vehicles for drug delivery, J. Control. Release 24 (1993) 119–132.
[13] P. Couvreur, C. Vauthier, Nanotechnology: intelligent design to treat complex disease, Pharm. Res. 23 (2006) 1417–1450.
[14] M.J. Alonso, Nanomedicines for overcoming biological barriers, Biomed. Pharmacother. 58 (2004) 168–172.
[15] A. Albanese, P.S. Tang, W.C.W. Chan, The effect of nanoparticle size, shape, and surface chemistry on biological systems, Annu. Rev. Biomed. Eng. 14 (2012) 1–16.
[16] H. Koo, M.S. Huh, I.C. Sun, S.H. Yuk, K. Choi, K. Kim, I.C. Kwon, In vivo targeted delivery of nanoparticles for theranosis, Acc. Chem. Res. 44 (2011) 1018–1028.
[17] R.A. Petros, J.M. DeSimone, Strategies in the design of nanoparticles for therapeutic applications, Nat. Rev. Drug Discov. 9 (2010) 615–627.
[18] S.A. Chew, S. Danti, Biomaterial-based implantable devices for cancer therapy, Adv. Healthc. Mater. 6 (2017).
[19] R. De Souza, P. Zahedi, C.J. Allen, M. Piquette-Miller, Polymeric drug delivery systems for localized cancer chemotherapy, Drug Deliv. 17 (2010) 365–375.
[20] J.B. Wolinsky, Y.L. Colson, M.W. Grinstaff, Local drug delivery strategies for cancer treatment: gels, nanoparticles, polymeric films, rods, and wafers, J. Control. Release 159 (2012) 14–26.
[21] E.A. Ho, P.L. Soo, C. Allen, M. Piquette-Miller, Impact of intraperitoneal, sustained delivery of paclitaxel on the expression of P-glycoprotein in ovarian tumors, J. Control. Release 117 (2007) 20–27.
[22] A.R.K. Sasikala, A.R. Unnithan, R.G. Thomas, S.W. Ko, Y.Y. Jeong, C.H. Park, C. S. Kim, Multifaceted implantable anticancer device for potential postsurgical breast cancer treatment: a single platform for synergistic inhibition of local regional breast cancer recurrence, surveillance, and healthy breast reconstruction, Adv. Funct. Mater. 28 (2018).
[23] A. GhavamiNejad, A.R.K. Sasikala, A.R. Unnithan, R.G. Thomas, Y.Y. Jeong, M. Vatankhah-Varnoosfaderani, F.J. Stadler, C.H. Park, C.S. Kim, Mussel-inspired electrospun smart magnetic nanofibers for hyperthermic chemotherapy, Adv. Funct. Mater. 25 (2015) 2867–2875.
[24] A.R.K. Sasikala, A.R. Unnithan, Y.H. Yun, C.H. Park, C.S. Kim, An implantable smart magnetic nanofiber device for endoscopic hyperthermia treatment and tumor-triggered controlled drug release, Acta Biomater. 31 (2016) 122–133.
[25] X.L. Xu, X.S. Chen, X.Y. Xu, T.C. Lu, X. Wang, L.X. Yang, X.B. Jing, BCNU-loaded PEG-PLLA ultrafine fibers and their in vitro antitumor activity against glioma C6 cells, J. Control. Release 114 (2006) 307–316.

[26] W.Y. Liu, J.C. Wei, Y.W. Chen, Electrospun poly(L-lactide) nanofibers loaded with paclitaxel and water-soluble fullerenes for drug delivery and bioimaging, New J. Chem. 38 (2014) 6223–6229.
[27] S.T. Yohe, V.L.M. Herrera, Y.L. Colson, M.W. Grinstaff, 3D superhydrophobic electrospun meshes as reinforcement materials for sustained local drug delivery against colorectal cancer cells, J. Control. Release 162 (2012) 92–101.
[28] Kenry, C.T. Lim, Nanofiber technology: current status and emerging developments, Prog. Polym. Sci. 70 (2017) 1–17.
[29] R. Vasita, D.S. Katti, Nanofibers and their applications in tissue engineering, Int. J. Nanomedicine 1 (2006) 15–30.
[30] J.J. Xue, J.W. Xie, W.Y. Liu, Y.N. Xia, Electrospun nanofibers: new concepts, materials, and applications, Acc. Chem. Res. 50 (2017) 1976–1987.
[31] G.R. Williams, N.P. Chatterton, T. Nazir, D.-G. Yu, L.-M. Zhu, C.J. Branford-White, Electrospun nanofibers in drug delivery: recent developments and perspectives, Ther. Deliv. 3 (2012) 515–533.
[32] A.R. Unnithan, A.R.K. Sasikala, P. Murugesan, M. Gurusamy, D.M. Wu, C.H. Park, C.S. Kim, Electrospun polyurethane-dextran nanofiber mats loaded with Estradiol for post-menopausal wound dressing, Int. J. Biol. Macromol. 77 (2015) 1–8.
[33] D. Li, Y.N. Xia, Electrospinning of nanofibers: reinventing the wheel? Adv. Mater. 16 (2004) 1151–1170.
[34] J. Doshi, D.H. Reneker, Electrospinning process and applications of electrospun fibers, J. Electrost. 35 (1995) 151–160.
[35] D. Kai, S.S. Liow, X.J. Loh, Biodegradable polymers for electrospinning: towards biomedical applications, Mat. Sci. Eng. C Mater. 45 (2014) 659–670.
[36] X.L. Hu, S. Liu, G.Y. Zhou, Y.B. Huang, Z.G. Xie, X.B. Jing, Electrospinning of polymeric nanofibers for drug delivery applications, J. Control. Release 185 (2014) 12–21.
[37] Y.J. Son, W.J. Kim, H.S. Yoo, Therapeutic applications of electrospun nanofibers for drug delivery systems, Arch. Pharm. Res. 37 (2014) 69–78.
[38] A. Balaji, M.V. Vellayappan, A.A. John, A.P. Subramanian, S.K. Jaganathan, E. Supriyanto, S.I.A. Razak, An insight on electrospun-nanofibers-inspired modern drug delivery system in the treatment of deadly cancers, RSC Adv. 5 (2015) 57984–58004.
[39] F. Masood, Polymeric nanoparticles for targeted drug delivery system for cancer therapy, Mat. Sci. Eng. C Mater. 60 (2016) 569–578.
[40] Z. Chen, Z.F. Chen, A.L. Zhang, J.M. Hu, X.M. Wang, Z.G. Yang, Electrospun nanofibers for cancer diagnosis and therapy, Biomater. Sci. 4 (2016) 922–932.
[41] Y.J. Kim, M. Ebara, T. Aoyagi, A smart hyperthermia nanofiber with switchable drug release for inducing cancer apoptosis, Adv. Funct. Mater. 23 (2013) 5753–5761.
[42] A.I. Baba, C. Câtoi, Comparative Oncology, The Publishing House of the Romanian Academy, Bucharest, 2007.
[43] V.V. Padma, An overview of targeted cancer therapy, Biomedicine 5 (2015) 1–6.
[44] S.M. Crafton, R. Salani, Beyond chemotherapy: an overview and review of targeted therapy in cervical cancer, Clin. Ther. 38 (2016) 449–458.
[45] G. Yang, J. Wang, Y. Wang, L. Li, X. Guo, S.B. Zhou, An implantable active-targeting micelle-in-nanofiber device for efficient and safe cancer therapy, ACS Nano 9 (2015) 1161–1174.
[46] Z.Y. Zhang, Y.J. Wu, G.Z. Kuang, S. Liu, D.F. Zhou, X.S. Chen, X.B. Jing, Y. B. Huang, Pt(IV) prodrug-backboned micelle and DCA loaded nanofibers for enhanced local cancer treatment, J. Mater. Chem. B 5 (2017) 2115–2125.
[47] S.H. Ranganath, C.H. Wang, Biodegradable microfiber implants delivering paclitaxel for post-surgical chemotherapy against malignant glioma, Biomaterials 29 (2008) 2996–3003.

[48] S. Liu, G.Y. Zhou, D.X. Liu, Z.G. Xie, Y.B. Huang, X. Wang, W.B. Wu, X.B. Jing, Inhibition of orthotopic secondary hepatic carcinoma in mice by doxorubicin-loaded electrospun polylactide nanofibers, J. Mater. Chem. B 1 (2013) 101–109.
[49] A.C. Palmer, P.K. Sorger, Combination cancer therapy can confer benefit via patient-to-patient variability without drug additivity or synergy, Cell 171 (2017) 1678.
[50] Y. Ma, X. Wang, S. Zong, Z.Y. Zhang, Z.G. Xie, Y.B. Huang, Y. Yue, S. Liu, X. B. Jing, Local, combination chemotherapy in prevention of cervical cancer recurrence after surgery by using nanofibers co-loaded with cisplatin and curcumin, RSC Adv. 5 (2015) 106325–106332.
[51] X.M. Luo, C.Y. Xie, H. Wang, C.Y. Liu, S.L. Yan, X.H. Li, Antitumor activities of emulsion electrospun fibers with core loading of hydroxycamptothecin via intratumoral implantation, Int. J. Pharm. 425 (2012) 19–28.
[52] U. Aggarwal, A.K. Goyal, G. Rath, Development and characterization of the cisplatin loaded nanofibers for the treatment of cervical cancer, Mat. Sci. Eng. C Mater. 75 (2017) 125–132.
[53] X.L. Xu, X.S. Chen, Z.F. Wang, X.B. Jing, Ultrafine PEG-PLA fibers loaded with both paclitaxel and doxorubicin hydrochloride and their in vitro cytotoxicity, Eur. J. Pharm. Biopharm. 72 (2009) 18–25.
[54] M.G. Ignatova, N.E. Manolova, R.A. Toshkova, I.B. Rashkov, E.G. Gardeva, L.S. Yossifova, M.T. Alexandrov, Electrospun nanofibrous mats containing quaternized chitosan and polylactide with in vitro antitumor activity against HeLa cells, Biomacromolecules 11 (2010) 1633–1645.
[55] S.U. Kumar, I. Matai, P. Dubey, B. Bhushan, A. Sachdev, P. Gopinath, Differentially cross-linkable core-shell nanofibers for tunable delivery of anticancer drugs: synthesis, characterization and their anticancer efficacy, RSC Adv. 4 (2014) 38263–38272.
[56] K.X. Qiu, C.L. He, W. Feng, W.Z. Wang, X.J. Zhou, Z.Q. Yin, L. Chen, H.S. Wang, X.M. Mo, Doxorubicin-loaded electrospun poly(L-lactic acid)/mesoporous silica nanoparticles composite nanofibers for potential postsurgical cancer treatment, J. Mater. Chem. B 1 (2013) 4601–4611.
[57] G.P. Ma, Y. Liu, C. Peng, D.W. Fang, B.J. He, J. Nie, Paclitaxel loaded electrospun porous nanofibers as mat potential application for chemotherapy against prostate cancer, Carbohydr. Polym. 86 (2011) 505–512.
[58] M. Leone, D.Y. Zhai, S. Sareth, S. Kitada, J.C. Reed, M. Pellecchia, Cancer prevention by tea polyphenols is linked to their direct inhibition of antiapoptotic Bcl-2-family proteins, Cancer Res. 63 (2003) 8118–8121.
[59] G.Y. Yang, J. Liao, K. Kim, E.J. Yurkow, C.S. Yang, Inhibition of growth and induction of apoptosis in human cancer cell lines by tea polyphenols, Carcinogenesis 19 (1998) 611–616.
[60] S. Okabe, M. Suganuma, M. Hayashi, E. Sueoka, A. Komori, H. Fujiki, Mechanisms of growth inhibition of human lung cancer cell line, PC-9, by tea polyphenols, Jpn. J. Cancer Res. 88 (1997) 639–643.
[61] S.J. Shao, L. Li, G. Yang, J.R. Li, C. Luo, T. Gong, S.B. Zhou, Controlled green tea polyphenols release from electrospun PCL/MWCNTs composite nanofibers, Int. J. Pharm. 421 (2011) 310–320.
[62] C.R. Pradeep, G. Kuttan, Effect of piperine on the inhibition of nitric oxide (NO) and TNF-alpha production, Immunopharmacol. Immunotoxicol. 25 (2003) 337–346.
[63] J. Wattanathorn, P. Chonpathompikunlert, S. Muchimapura, A. Priprem, O. Tankamnerdthai, Piperine, the potential functional food for mood and cognitive disorders, Food Chem. Toxicol. 46 (2008) 3106–3110.
[64] J.S. Bang, D.H. Oh, H.M. Choi, B.J. Sur, S.J. Lim, J.Y. Kim, H.I. Yang, M.C. Yoo, D. H. Hahm, K.S. Kim, Anti-inflammatory and antiarthritic effects of piperine in human interleukin 1 beta-stimulated fibroblast-like synoviocytes and in rat arthritis models, Arthritis Res. Ther. 11 (2009).

[65] S. Jain, S.R.K. Meka, K. Chatterjee, Engineering a piperine eluting nanofibrous patch for cancer treatment, ACS Biomater. Sci. Eng. 2 (2016) 1376–1385.
[66] N. Chainani-Wu, Safety and anti-inflammatory activity of curcumin: a component of tumeric (Curcuma longa), J. Altern. Complement. Med. 9 (2003) 161–168.
[67] R. Kuttan, P. Bhanumathy, K. Nirmala, M.C. George, Potential anticancer activity of turmeric (Curcuma longa), Cancer Lett. 29 (1985) 197–202.
[68] R. De, P. Kundu, S. Swarnakar, T. Ramamurthy, A. Chowdhury, G.B. Nair, A.K. Mukhopadhyay, Antimicrobial activity of curcumin against Helicobacter pylori isolates from India and during infections in mice, Antimicrob. Agents Chemother. 53 (2009) 1592–1597.
[69] R. Sedghi, A. Shaabani, Z. Mohammadi, F.Y. Samadi, E. Isaei, Biocompatible electrospinning chitosan nanofibers: a novel delivery system with superior local cancer therapy, Carbohydr. Polym. 159 (2017) 1–10.
[70] R.J. Williams, J.P.E. Spencer, C. Rice-Evans, Flavonoids: antioxidants or signalling molecules? Free Radic. Biol. Med. 36 (2004) 838–849.
[71] D.F. Birt, S. Hendrich, W.Q. Wang, Dietary agents in cancer prevention: flavonoids and isoflavonoids, Pharmacol. Ther. 90 (2001) 157–177.
[72] A. Hervault, N.T.K. Thanh, Magnetic nanoparticle-based therapeutic agents for thermo-chemotherapy treatment of cancer, Nanoscale 6 (2014) 11553–11573.
[73] J.L. Li, X.L. Hou, H.C. Bao, L. Sun, B. Tang, J.F. Wang, X.G. Wang, M. Gu, Graphene oxide nanoparticles for enhanced photothermal cancer cell therapy under the irradiation of a femtosecond laser beam, J. Biomed. Mater. Res. A 102 (2014) 2181–2188.
[74] D. Jaque, L.M. Maestro, B. del Rosal, P. Haro-Gonzalez, A. Benayas, J.L. Plaza, E. M. Rodriguez, J.G. Sole, Nanoparticles for photothermal therapies, Nanoscale 6 (2014) 9494–9530.
[75] Z. Zhang, S. Liu, H. Xiong, X. Jing, Z. Xie, X. Chen, Y. Huang, Electrospun PLA/MWCNTs composite nanofibers for combined chemo- and photothermal therapy, Acta Biomater. 26 (2015) 115–123.
[76] X.C. Wang, F. Lv, T. Li, Y.M. Han, Z.F. Yi, M.Y. Liu, J. Chang, C.T. Wu, Electrospun micropatterned nanocomposites incorporated with Cu2S nanoflowers for skin tumor therapy and wound healing, ACS Nano 11 (2017) 11337–11349.

CHAPTER 3

Transdermal drug delivery via microneedle patches

Amin GhavamiNejad, Brian Lu, Xiao Yu Wu
Advanced Pharmaceutics and Drug Delivery Laboratory, Leslie L. Dan Faculty of Pharmacy, University of Toronto, Toronto, ON, Canada

Contents

3.1 Introduction

Transdermal drug delivery systems using noninvasive or minimally invasive technologies are an important part of novel drug delivery systems [1, 2]. Among these systems, microneedles (MNs) have garnered considerable attention in recent years, particularly for protein and vaccine delivery to treat diseases such as diabetes and cancer [3–5]. The MNs are three-dimensional microstructures with lengths usually $<1000\ \mu m$ [6]. The MNs can be designed with different needle lengths to reach an appropriate depth without harming vascular networks and affecting nerves of the skin. As such, MNs can be used to transcutaneously deliver both hydrophilic and hydrophobic drugs and therapeutic biomolecules, effectively overcoming the barrier of stratum corneum that limits the efficiency of conventional transdermal drug delivery [7].

Metallic MNs were the first type of MNs developed for transdermal use. However, a low drug-loading capability, potential breakdown in the skin, and costly fabrication methods have greatly restricted their clinical application [8]. To overcome these issues, polymers have been utilized to fabricate advanced MNs for drug delivery applications [9, 10]. Polymeric MNs can be designed to have controllable drug release rates over long periods of time.

Biomimetic Nanoengineered Materials for Advanced Drug Delivery
https://doi.org/10.1016/B978-0-12-814944-7.00003-5

Many studies have demonstrated that polymeric MNs can effectively deliver both low and high molecular weight drugs including heparin [11], gentamicin [12], lidocaine [13], donepezil [14], proteins such as insulin [15], and vaccines for influenza [16], ebola [17], hepatitis B [18], and other antigens [19] for the treatment and prevention of a variety of diseases. Most recently, MNs have been used for intracellular delivery of siRNA [20], DNA [21], and other therapeutic biomolecules for cancer treatments. This chapter summarizes the recent advances in the development of polymeric MNs for the treatment of diabetes and cancer, the two of leading diseases in humans.

3.2 Drug delivery via MN patches for diabetes treatment

Diabetes is a growing epidemic with an estimated 8.5% global prevalence rate [22]. In 2014, >422 million adults had diabetes, nearly a fourfold increase from 108 million in 1980 [23]. Of all the people with diabetes, 10% are type 1 diabetic (T1D), who require lifetime treatment with insulin to sustain their lives and maintain their glycemic targets. To deliver insulin on demand at high glucose levels, several insulin delivery systems have been developed, including glucose-responsive implantable devices [24–28] and MN patches [29–31]. In light of a lower risk in potential insulin overdose than implants, MN patches for insulin delivery have gained increasing interests and are currently being developed by companies for patients who would otherwise require painful injections of exogenous insulin to maintain normal glucose levels.

Recently, "smart" (glucose-responsive) insulin MN patch systems have been designed to transdermally deliver insulin for hyperglycemia (high blood sugar level) [30, 31]. The Gu group designed an MN array patch containing glucose-sensitive vesicles that were loaded with glucose oxidase (GOx) and insulin (Fig. 3.1). The vesicles dissociated in the presence of excess H_2O_2, a by-product of glucose oxidation catalyzed by the co-loaded GOx. As a result, more insulin was liberated from the vesicles at higher glucose concentration and released from the MN system to the body. This composite MN system not only exhibited rapid response to physiologically relevant glucose stimuli, but also reduced the risk of hypoglycemia due to its ability to halt insulin delivery at low glucose levels.

In addition to glucose-responsive MN design, dissolving MN arrays were developed for transdermal insulin delivery. A two-layered patch with dissolving MNs was fabricated by Lee et al. [31] Insulin-loaded dissolvable MNs were made from polyvinylpyrrolidone (PVP) of two different

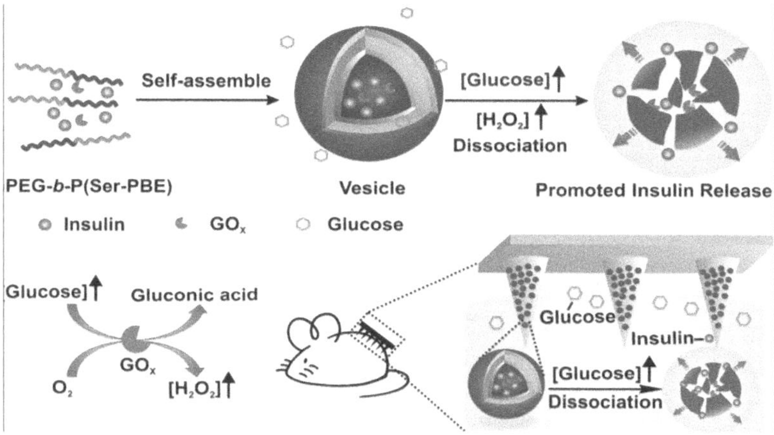

Fig. 3.1 Glucose-responsive insulin- and GOx-loaded vesicles that are designed to dissociate in the presence of increased levels of H_2O_2 corresponding to high glucose levels. The dissociation of the vesicles promotes insulin release at high glucose levels. The vesicles were loaded into hyaluronic acid (HA)-based MNs for glucose-responsive transdermal insulin delivery in a T1D mouse model. *(Reprinted from X.L. Hu, J.C. Yu, C.G. Qian, Y. Lu, A.R. Kahkoska, Z.G. Xie, X.B. Jing, J.B. Buse, Z. Gu, H_2O_2-responsive vesicles integrated with transcutaneous patches for glucose-mediated insulin delivery, ACS Nano 11 (2017) 613–620. Copyright 2010, with permission from American Chemical Society).*

molecular weights at an optimal ratio of PVP10/PVP360 determined on the mechanical properties of the needles (Fig. 3.2). The MNs successfully penetrated excised porcine skin with no indication of mechanical failure. The MNs rapidly dissolved in the skin after insertion, resulting in a rapid release of loaded insulin, thereby regulating the glucose level up to 4 h posttreatment in diabetic mouse model.

A swellable polyvinyl alcohol (PVA)-based MN patch was designed by Jin and coworkers for transdermal delivery of insulin [32]. Based on the property of temperature-dependent phase transition of PVA, the group used freeze-thaw treatment to induce microcrystalline cross-linking in the MNs to reinforce the MNs (Fig. 3.3). The mild physical cross-linking condition is friendly to protein drugs and allows easy loading of insulin. The needle tips were sufficiently stiff at dry state and readily swelled by the interstitial fluid upon skin penetration. The swelling facilitated the release of loaded insulin while the microcrystalline cross-linkages maintained the integrity of hydrated needle tips, allowing the complete removal of intact needles.

A live pancreatic β-cell loaded with composite MN patch was devised by the Gu group for glucose-responsive insulin secretion and transdermal

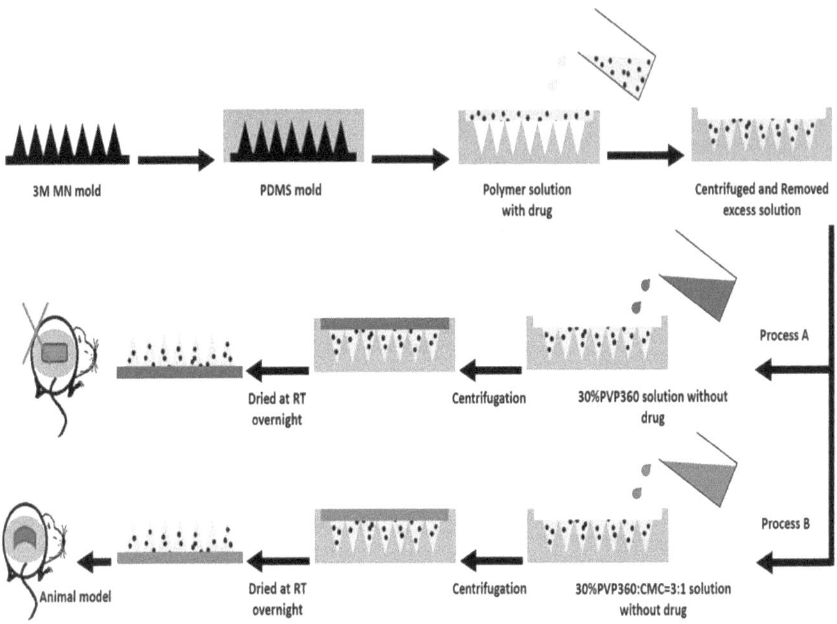

Fig. 3.2 Two-step fabrication process of dual-layer PVP360/PVP10 MN patches loaded with drug (e.g., insulin). Process A uses PVP360 polymer solution as the backing layer for the MN patch. Process B uses PVP360/CMC polymer mixture as the backing layer for the MN patch that can be used for animal study. *(Reprinted from I.C. Lee, Y.C. Wu, S.W. Tsai, C.H. Chen, M.H. Wu, Fabrication of two-layer dissolving polyvinylpyrrolidone microneedles with different molecular weights for in vivo insulin transdermal delivery, RSC Adv. 7 (2017) 5067–5075. Copyright 2017, with permission from Royal society of Chemistry).*

delivery during hyperglycemia [30]. The cells were encapsulated in cross-linked gels which were then incorporated at the base of the MN patch made of cross-linked hyaluronic acid (HA). To enhance glucose transport and sensing, the MN tips were loaded with amylose and synthetic "glucose-signal amplifiers" (GSAs) microgels (Fig. 3.4). The GSA microgels were prepared from hypoxia-sensitive alginate, GOx, α-amylase (AM), and glucoamylase (GA). At a high glucose concentration, the oxygen level in the microenvironment reduces due to the enzymatic oxidation of glucose in the presence of GOx. The depletion of oxygen caused dissociation of the GSA microgel and release of AM which in turn broke down the embedded amylose near the base of the MN into smaller oligosaccharides that were further converted to glucose molecules by GA. The glucose molecules in turn trigger the β-cell capsules at the base of the MN, promoting insulin release

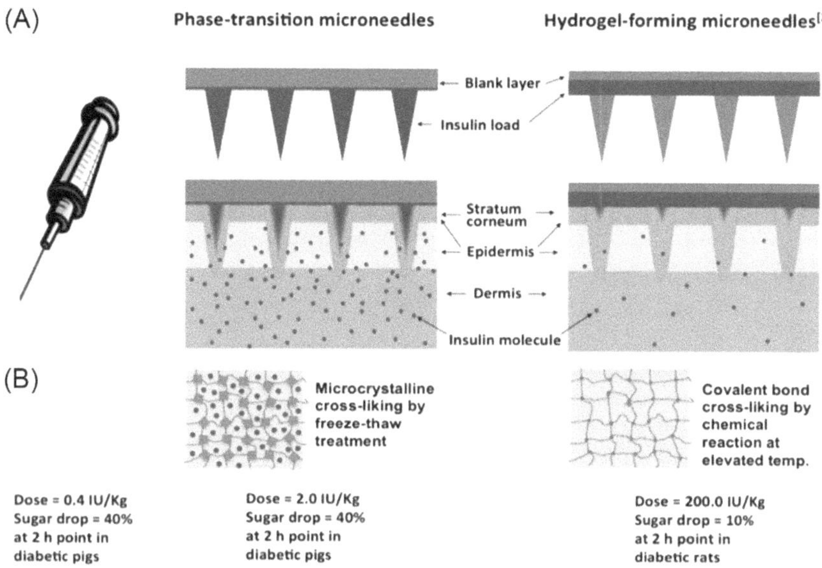

Fig. 3.3 Drug release mechanism of Phase-Transition MN (PTM) patches. (A) Swelling of microneedles due to the absorption of body fluids that provide insulin delivery in the dermis layer. (B) The advantage of freeze-thaw crosslinked PTM over covalently crosslinked hydrogel-forming MNs (HFM). Microcrystalline domains in the PTM allow for direct insulin loading into the needle tips, which results in higher levels of insulin delivered compared to HFM. *(Reprinted from S.X. Yang, F. Wu, J.G. Liu, G.R. Fan, W. Welsh, H. Zhu, T. Jin, Phase-transition microneedle patches for efficient and accurate transdermal delivery of insulin, Adv. Funct. Mater. 25 (2015) 4633–4641, Copyright 2015, with permission from Wiley).*

through the needle tips into the body. This sophisticated design may address the problems of long-term efficacy and immune response associated with β-cell therapy. Yet maintaining the viability of β-cells and activity of multiple enzymes during the fabrication and storage of the MN patch presents another kind of challenges. Another dissolvable MN patch was reported by Ling et al., which was made from starch and gelatin and released encapsulated insulin upon dissolution (Fig. 3.5) [33]. The MN patch was able to load insulin without changing its secondary structure and to provide good stability at room and body temperature. After application of the insulin-loaded MN patch onto the shaven backside of diabetic rats, the time for reaching the minimum plasma glucose level was longer than that of the insulin injected control group without a sign of hypoglycemia. These results showed that the dissolvable MNs enabled loading and delivery of unstable biomolecules.

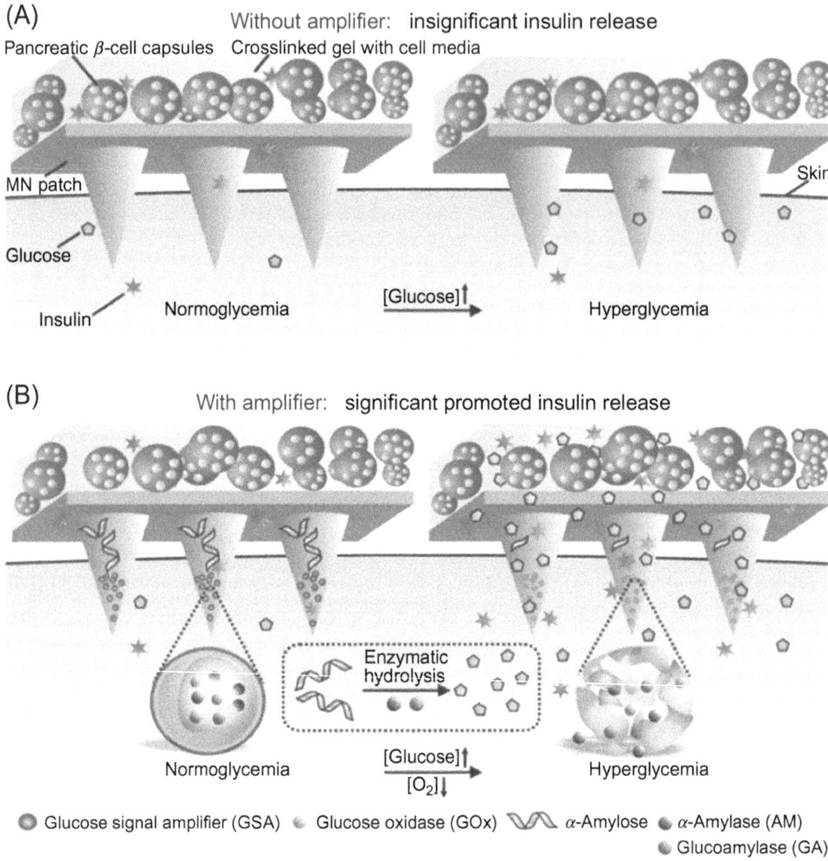

Fig. 3.4 Schematic illustration of MN patch containing pancreatic β-cells and glucose signal amplifiers (GSA). (A) In the absence of GSA, endogenous glucose molecules cannot reach the β-cells at both euglycemic and hyperglycemic conditions and therefore resulting in an insignificant delivery of insulin. (B) In the presence of amylose and GSA, sufficient insulin release is achieved at hyperglycemia due to the amplification of glucose signal to the β-cells. *(Reprinted from Y.Q. Ye, J.C. Yu, C. Wang, N.Y. Nguyen, G.M. Walker, J.B. Buse, Z. Gu, Microneedles integrated with pancreatic cells and synthetic glucose-signal amplifiers for smart insulin delivery, Adv. Mater. 28 (2016) 3115–3121, Copyright 2016, with permission from Wiley).*

In contrast to extensive studies on transdermal delivery of insulin, little research has been conducted until very recently on the transdermal delivery of glucagon or other agents for raising blood sugar level at hypoglycemia. Given the fact that a major and potentially fatal complication of insulin therapy in diabetes is hypoglycemia (low blood sugar), prevention of hypoglycemia is a significant unmet medical need. To address excessive

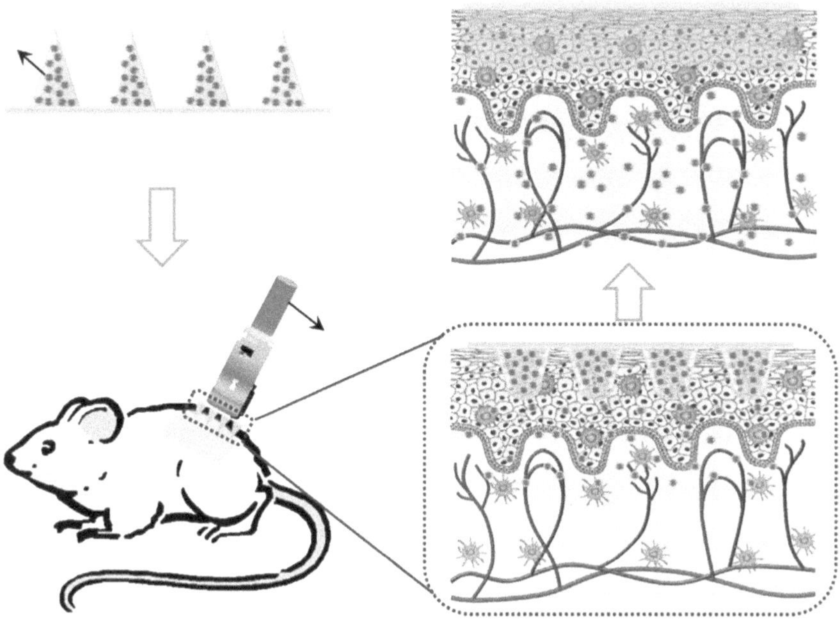

Fig. 3.5 Schematic of the application of insulin-loaded dissolvable starch/gelatin MN patch, which can rapidly dissolve in the skin to release preloaded insulin. *(Reprinted from M.H. Ling, M.C. Chen, Dissolving polymer microneedle patches for rapid and efficient transdermal delivery of insulin to diabetic rats, Acta. Biomater. 9 (2013) 8952–8961, Copyright 2013, with permission from Elsevier).*

insulin-induced hypoglycemia, Gu and coworkers recently introduced an MN patch to deliver glucagon in response to insulin stimuli [34]. In this work, an insulin aptamer, with a single stranded oligonucleotide, was conjugated to glucagon molecule. The glucagon conjugates were then mixed with immobilized insulin and loaded into an HA-based MN patch. In the presence of a high insulin concentration, which represented excessive insulin-induced hypoglycemia, the glucagon conjugates were released from the MNs due to the displacement of the immobilized insulin-glucagon conjugate with free insulin molecules (Fig. 3.6). The authors further demonstrated that the transdermal application of the patch prevented insulin-induced hypoglycemia in a diabetic mouse model. Despite the impressive results, an insulin-responsive glucagon delivery system would not counteract other possible causes of hypoglycemia than excessive insulin, such as overexercise, missed meals, or excessive consumption of alcohol that can impair the liver from liberating glucose when needed, which may occur

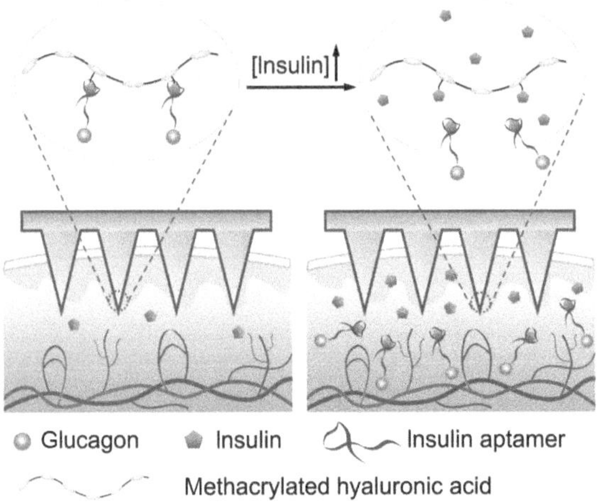

Fig. 3.6 Insulin-responsive glucagon-conjugate release mechanism of the composite MN patch. *(Reprinted from J.C. Yu, Y.Q. Zhang, W.J. Sun, A.R. Kahkoska, J.Q. Wang, J.B. Buse, Z. Gu, Insulin-responsive glucagon delivery for prevention of hypoglycemia, Small, 13 (2017). Copyright 2017, with permission from Wiley).*

even at low concentrations of insulin. Moreover, sensitivity to insulin can vary significantly among patients, especially in those with insulin resistance. All these scenarios could render the presetting insulin target difficult.

To prevent the occurrence of hypoglycemia in a broad range of scenarios, Wu and coworkers proposed a glucose-responsive glucagon-MN patch. This system is composed of glucose-responsive zwitterionic polymer-based microgels with encapsulated glucogan, which were then incorporated into methacrylated hyaluronic acid (MeHA)-based MNs. The zwitterionic monomer unit stabilized glucagon due to its excellent protein stabilization property [35–39] while boronic acid unit provided hypoglycemia-triggered glucagon release via its mono-complexation to bis-complexation with glucose [40, 41]. The photo-cross-linked composite MNs were strong enough to penetrate rat skin and swelled quickly in the interstitial fluid. When applied to T1D rats, the glucagon-loaded MN patch was able to prevent the onset of hypoglycemia after the administration of an overdose of insulin.

3.3 Drug delivery via MN patches for cancer treatment

Among all the major diseases, cancer remains the leading cause of death worldwide. Recently, a few groups have explored MN technologies for cancer therapies. Owing to the limitation of tumor accessibility of MNs,

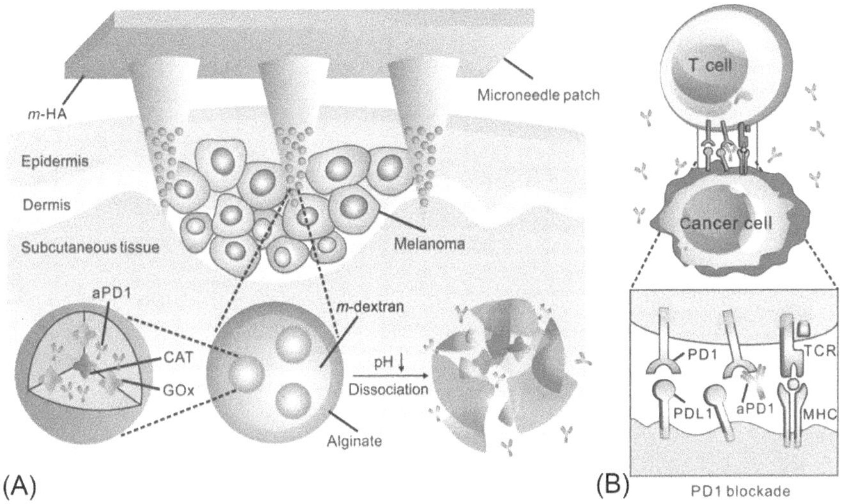

Fig. 3.7 Schematic representation of aPD1 delivery via HA-based MN patch. (A) Enzymatic oxidation of glucose to gluconic acid provides an acidic environment that enhances the dissociation of the pH-sensitive particles for the rapid delivery of aPD1. (B) Mechanism of aPD1 to activate the immune system for the destruction of melanoma cells. *(Reprinted from C. Wang, Y.Q. Ye, G.M. Hochu, H. Sadeghifar, Z. Gu, Enhanced cancer immunotherapy by microneedle patch-assisted delivery of anti-PD1 antibody, Nano Lett. 16 (2016) 2334–2340. Copyright 2016, with permission from American Chemical Society).*

MN technologies have been attempted only for treatment of skin and oral cancers. Wang et al. designed an MN-based system for local delivery of anti-PD-1 (aPD1) antibodies for skin cancer treatment (Fig. 3.7) [42]. The MN was made of HA-based polymer loaded with pH-responsive dextran, aPD1, and GOx/catalase (CAT). This system was operated based on the enzymatic oxidation of glucose to gluconic acid in the presence of oxygen, which produced an acidic environment that enhances the dissociation of pH-sensitive particles for the rapid release of aPD1. The authors found that GOx-loaded MN patch, MN-GOx aPD1, delivered more aPD1 to the tumor tissue and generated significantly higher antitumor efficacy than intratumoral injection of free aPD1 or MN-aPD1 without GOx.

In another publication, Chen et al. reported a smart composite MN system with controllable drug release profile and heat generation under near-infrared (NIR) radiation, for chemophotothermal cancer therapy [43]. In this system, photosensitive lanthanum hexaboride (LaB_6) and an anticancer drug (doxorubicin; DOX) were incorporated into a polycaprolactone MN

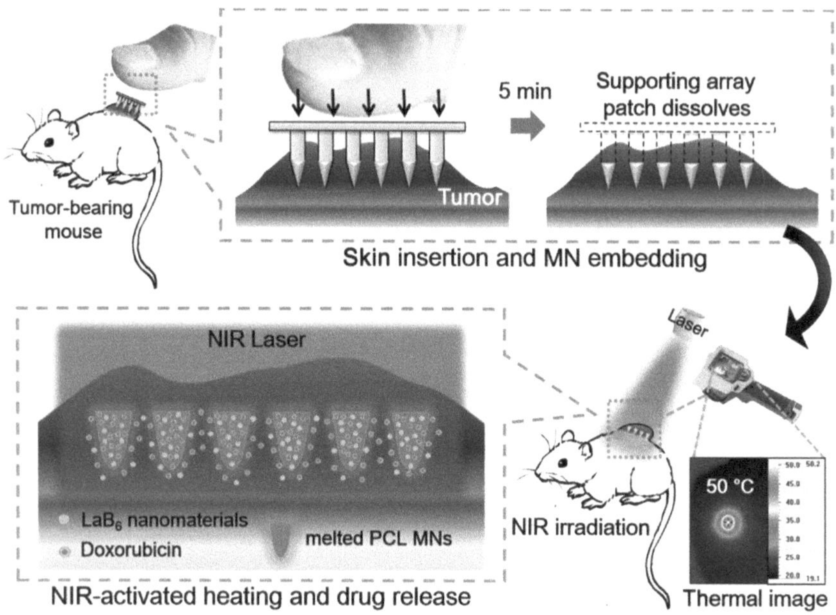

Fig. 3.8 Schematic representation of NIR-induced chemophotothermal skin cancer therapy via MN patch. The dissolution of the PVA/PVP backing layers leave the DOX- and LaB_6-loaded MN tips intact under the skin. Upon NIR irradiation, the MN tips can convert the NIR light to heat and facilitate the release of DOX to kill the cancer cells through photothermal ablation and chemotherapy, respectively. *(Reprinted from C.M.C. Chen, Z.W. Lin, M.H. Ling, Near-infrared light-activatable microneedle system for treating superficial tumors by combination of chemotherapy and photothermal therapy, ACS Nano, 10 (2016) 93–101. Copyright 2016, with permission from American Chemical Society).*

system with a dissolvable PVA/PVP supporting patch matrix. When NIR was applied to the MN, the light was absorbed and dissipated locally as heat to induce local hyperthermia to destroy the cancer cells (Fig. 3.8). It was shown that the release of DOX could be controlled by NIR light on/off for several cycles. In vivo studies have shown that the chemophotothermal treatment with the NIR-sensitive composite MN system completely eradicated 4T1 murine breast tumors within 7 days without tumor recurrence and body weight loss in mice.

Similarly, Hao et al. fabricated an NIR light-sensitive poly(L-lactide) (PLLA) MN system coated with PEGylated gold nanorods (PEG-GNRs) for treating A431 human skin tumor in mice (Fig. 3.9) [44]. The as-made MNs generated heat by NIR light; which then effectively transmitted to the middle of the tumor sites ablating cancer cells via photothermal effect.

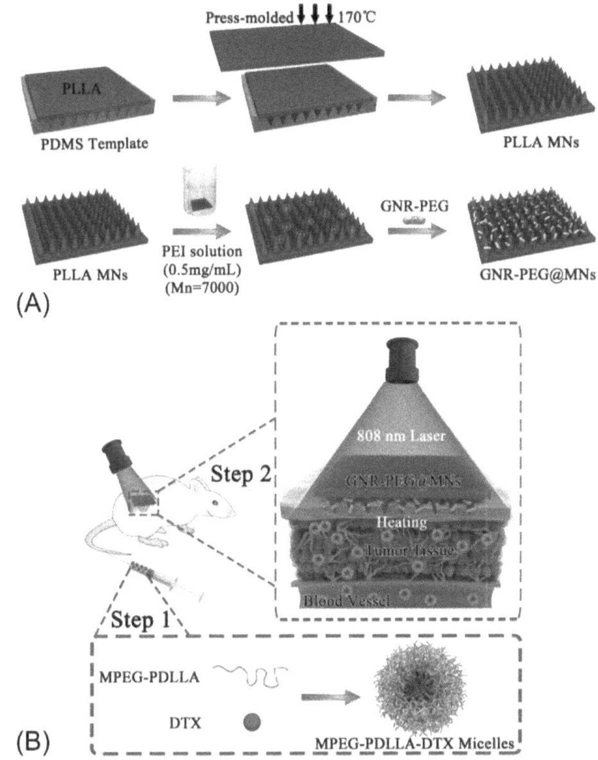

Fig. 3.9 (A) Fabrication process of PLLA-based MN patch and loading with GNR-PEG. (B) Chemophotothermal cancer therapy mediated by pre-injected DTX-loaded micelles and NIR laser irradiation to the GNR-PEG-loaded MN patch. *(Reprinted from Y. Hao, M.L. Dong, T.Y. Zhang, J.R. Peng, Y.P. Jia, Y.P. Cao, Z.Y. Qian, Novel approach of using near-infrared responsive PEGylated gold nanorod coated poly(L-lactide) microneedles to enhance the antitumor efficiency of docetaxel-loaded MPEG-PDLLA micelles for treating an A431 Tumor, ACS Appl. Mater. Interfaces, 9 (2017) 15317–15327. Copyright 2017, with permission from American Chemical Society).*

In combination with chemotherapy using docetaxel (DTX)-loaded micelles, this MN system was shown to completely destroy the A431 tumor.

The same strategy was also used by Ye et al. who prepared a tumor lysate (B16F10) containing melanin and loaded the mixture into HA-based MNs to deliver the lysate in an NIR-controllable manner [45]. The group used melanin as a photothermal agent to convert light into heat for photothermal ablation [46]. They investigated the temperature variations of the MNs by exposing the samples to 808-nm NIR laser at a power density of 1 W/cm^2. The group observed significant and rapid heating by MNs loaded with the tumor lysate. The produced heat promoted tumor-antigen uptake by dendritic cells, resulting in an improvement in the treatment. This method delayed the growth of distant tumor and improved long-term survival in prophylactic models.

Lee et al. reported a dissolving MN patch for the delivery of human growth hormone (hGH) for cancer treatment [47]. In this study, the hGH was loaded into the tips of dissolving MNs composed of carboxymethylcellulose (CMC) and trehalose via a mild process to avoid loss of bioactivity of the hGH. The pharmacokinetics of hGH delivered by the MNs via the rat skin was similar to the control groups treated with subcutaneous (SC) injection (Fig. 3.10). After patch application, the needle tips were almost completely dissolved. The hGH retained its pharmacological activity after encapsulation and release from the MNs. This MN formulation maintained the stability and bioactivity of hGH at room temperature for more than

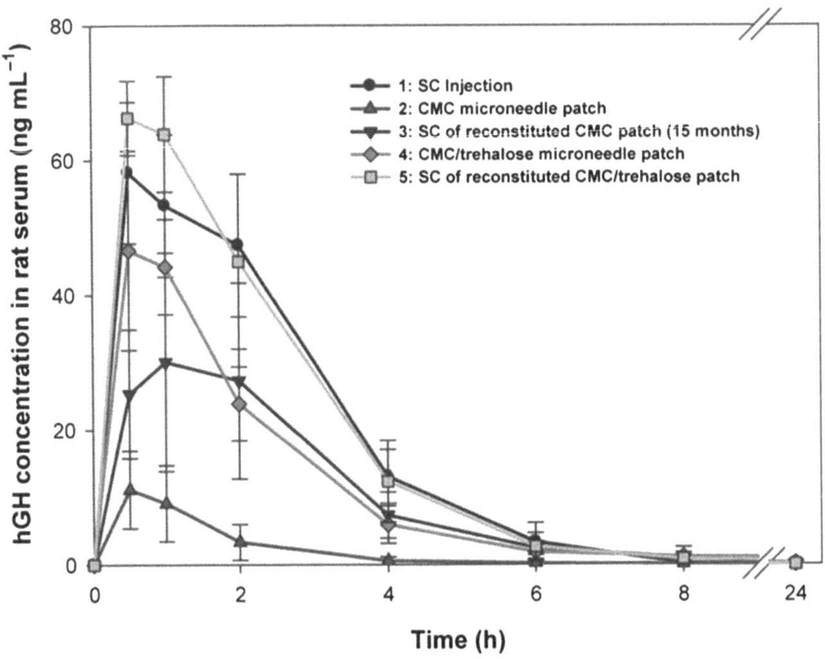

Fig. 3.10 hGH concentration in rat serum after MN patch application. Released hGH pharmacokinetics from MN had a profile similar to conventional subcutaneous injection. Peak hGH plasma levels occurred at 0.7 h. *(Reprinted from J.W. Lee, S.O. Choi, E.I. Felner, M.R. Prausnitz, Dissolving microneedle patch for transdermal delivery of human growth hormone, Small, 7 (2011) 531–539. Copyright 2011, with permission from Wiley).*

1 year. In vivo results show that hGH released from the patch exhibits bioavailability similar to the SC-injected hGH (control group), indicating a potential of this MN system for protein delivery.

3.4 Conclusion

This chapter provides a short review about some of the latest advances in the development of various polymeric MN patches for the treatment of diabetes and cancer. MN patches have been widely investigated for transdermal delivery of proteins and peptides, such as insulin and glucagon for diabetes treatment and hGH for cancer treatment. To preserve the native structure and pharmacological activity of protein and peptide drugs, avoidance of heat and high-energy light exposure during MN fabrication is necessary. To achieve high efficiency of transdermal delivery of biomolecules, the geometry of polymeric MNs needs to be optimized, for example, to possess sharper tips and smaller aspect ratios that exhibit better skin insertion. Multifunctional MNs have found useful applications in combination therapies of cancer, such as chemophotothermal therapy. The excellent properties of thermal generation by radiation-sensitive composite MNs can be exploited and combined with chemotherapy to obtain a synergistic effect. Owing to the limited penetration depth of needles, nevertheless, cancer therapy via MNs may only be applicable to certain types of cancers, such as skin and oral cancers.

Acknowledgments

This work was supported by a grant from JDRF (grant no. 2-SRA-2016-268-A-N). The BBDC Seller Fellowship to A.G and the OGS scholarship to B.L. are also gratefully acknowledged.

References

[1] M.N. Pastore, Y.N. Kalia, M. Horstmann, M.S. Roberts, Transdermal patches: history, development and pharmacology, Br. J. Pharmacol. 172 (9) (2015) 2179–2209.
[2] K. van der Maaden, W. Jiskoot, J. Bouwstra, Microneedle technologies for (trans)dermal drug and vaccine delivery, J. Control. Release 161 (2) (2012) 645–655.
[3] R. Donnelly, D. Douroumis, Microneedles for drug and vaccine delivery and patient monitoring, Drug Deliv. Transl. Res. 5 (4) (2015) 311–312.
[4] H.L. Quinn, M.C. Kearney, A.J. Courtenay, M.T.C. McCrudden, R.F. Donnelly, The role of microneedles for drug and vaccine delivery, Expert Opin. Drug Deliv. 11 (11) (2014) 1769–1780.

[5] Y.C. Kim, J.H. Park, M.R. Prausnitz, Microneedles for drug and vaccine delivery, Adv. Drug Deliv. Rev. 64 (14) (2012) 1547–1568.
[6] K. Moronkeji, S. Todd, I. Dawidowska, S.D. Barrett, R. Akhtar, The role of subcutaneous tissue stiffness on microneedle performance in a representative in vitro model of skin, J. Control. Release 265 (2017) 102–112.
[7] S.S. Morteza Amjadi, B.J. Nelson, M. Sitti, Recent advances in wearable transdermal delivery systems, Adv. Mater. (2018).
[8] R.F. Donnelly, T.R.R. Singh, A.D. Woolfson, Microneedle-based drug delivery systems: microfabrication, drug delivery, and safety, Drug Deliv. 17 (4) (2010) 187–207.
[9] M. Wang, L.Z. Hu, C.J. Xu, Recent advances in the design of polymeric microneedles for transdermal drug delivery and biosensing, Lab Chip 17 (8) (2017) 1373–1387.
[10] A. GhavamiNejad, C.H. Park, C.S. Kim, In situ synthesis of antimicrobial silver nanoparticles within antifouling zwitterionic hydrogels by catecholic redox chemistry for wound healing application, Biomacromolecules 17 (3) (2016) 1213–1223.
[11] Y.Q. Zhang, J.C. Yu, J.Q. Wang, N.J. Hanne, Z. Cui, C.G. Qian, C. Wang, H.L. Xin, J.H. Cole, C.M. Gallippi, Y. Zhu, Z. Gu, Thrombin-responsive transcutaneous patch for auto-anticoagulant regulation, Adv. Mater. 29 (4) (2017).
[12] P. Gonzalez-Vazquez, E. Larraneta, M.T.C. McCrudden, C. Jarrahian, A. Rein-Weston, M. Quintanar-Solares, D. Zehrung, H. McCarthy, A.J. Courtenay, R.F. Donnelly, Transdermal delivery of gentamicin using dissolving microneedle arrays for potential treatment of neonatal sepsis, J. Control. Release 265 (2017) 30–40.
[13] X. Xie, C. Pascual, C. Lieu, S. Oh, J. Wang, B.D. Zou, J.L. Xie, Z.H. Li, J. Xie, D.C. Yeomans, M.X. Wu, X.S. Xie, Analgesic microneedle patch for neuropathic pain therapy, ACS Nano 11 (1) (2017) 395–406.
[14] M.C. Kearney, E. Caffarel-Salvador, S.J. Fallows, H.O. McCarthy, R.F. Donnelly, Microneedle-mediated delivery of donepezil: potential for improved treatment options in Alzheimer's disease, Eur. J. Pharm. Biopharm. 103 (2016) 43–50.
[15] R.J. Pettis, B. Ginsberg, L. Hirsch, D. Sutter, S. Keith, E. McVey, N.G. Harvey, M. Hompesch, L. Nosek, C. Kapitza, L. Heinemann, Intradermal microneedle delivery of insulin lispro achieves faster insulin absorption and insulin action than subcutaneous injection, Diabetes Technol. Ther. 13 (4) (2011) 435–442.
[16] N.G. Rouphael, M.J. Mulligan, Microneedle patch for immunization of immunocompromised hosts, Oncotarget 8 (55) (2017) 93311–93312.
[17] H.W. Yang, L. Ye, X.D. Guo, C.L. Yang, R.W. Compans, M.R. Prausnitz, Ebola vaccination using a DNA vaccine coated on PLGA-PLL/gamma PGA nanoparticles administered using a microneedle patch, Adv Healthc Mater 6 (1) (2017).
[18] D. Poirier, F. Renaud, V. Dewar, L. Strodiot, F. Wauters, J. Janimak, T. Shimada, T. Nomura, K. Kabata, K. Kuruma, T. Kusano, M. Sakai, H. Nagasaki, T. Oyamada, Hepatitis B surface antigen incorporated in dissolvable microneedle array patch is antigenic and thermostable, Biomaterials 145 (2017) 256–265.
[19] K. Matsuo, Y. Yokota, Y. Zhai, Y.S. Quan, F. Kamiyama, Y. Mukai, N. Okada, S. Nakagawa, A low-invasive and effective transcutaneous immunization system using a novel dissolving microneedle array for soluble and particulate antigens, J. Control. Release 161 (2012) 10. 2014, 184, 9-9.
[20] Y. Deng, J. Chen, Y. Zhao, X.H. Yan, L. Zhang, K.W. Choy, J. Hu, H.J. Sant, B.K. Gale, T. Tang, Transdermal delivery of siRNA through microneedle array, Sci. Rep. 6 (2016).
[21] Y. Lee, S.R. Dugansani, S.H. Jeon, S.H. Hwang, J.H. Kim, S.H. Park, J.H. Jeong, Drug-delivery system based on salmon DNA nano- and micro-scale structures, Sci. Rep. 7 (2017).
[22] K. Rice, B.T. Hiwi, M. Zwarenstein, B. Lavallee, D.E. Barre, S.B. Harris, F.A.P. Team, Best practices for the prevention and management of diabetes and

obesity-related chronic disease among indigenous peoples in Canada: a review, Can. J. Diabetes 40 (3) (2016) 216–225.
[23] P.E.H. Schwarz, G. Gallein, D. Ebermann, A. Muller, A. Lindner, U. Rothe, I.T. Nebel, G. Muller, Global diabetes survey—an annual report on quality of diabetes care, Diabetes Res. Clin. Pract. 100 (1) (2013) 11–18.
[24] C.R. Gordijo, K. Koulajian, A.J. Shuhendler, L.D. Bonifacio, H.Y. Huang, S. Chiang, G.A. Ozin, A. Giacca, X.Y. Wu, Nanotechnology-enabled closed loop insulin delivery device: in vitro and in vivo evaluation of glucose-regulated insulin release for diabetes control, Adv. Funct. Mater. 21 (1) (2011) 73–82.
[25] J. Li, M.K. Chu, B. Lu, S. Mirzaie, K. Chen, C.R. Gordijo, O. Plettenburg, A. Giacca, X.Y. Wu, Enhancing thermal stability of a highly concentrated insulin formulation with Pluronic F-127 for long-term use in microfabricated implantable devices, Drug Deliv. Transl. Res. 7 (4) (2017) 529–543.
[26] M.K.L. Chu, J. Chen, C.R. Gordijo, S. Chiang, A. Ivovic, K. Koulajian, A. Giacca, X.Y. Wu, Y. Sun, In vitro and in vivo testing of glucose-responsive insulin-delivery microdevices in diabetic rats, Lab Chip 12 (14) (2012) 2533–2539.
[27] M.K. Chu, C.R. Gordijo, J. Li, A.Z. Abbasi, A. Giacca, O. Plettenburg, X.Y. Wu, In vivo performance and biocompatibility of a subcutaneous implant for real-time glucose-responsive insulin delivery, Diabetes Technol. Ther. 17 (4) (2015) 255–267.
[28] C.R. Gordijo, A.J. Shuhendler, X.Y. Wu, Glucose-responsive bioinorganic nanohybrid membrane for self-regulated insulin release, Adv. Funct. Mater. 20 (9) (2010) 1404.
[29] X.L. Hu, J.C. Yu, C.G. Qian, Y. Lu, A.R. Kahkoska, Z.G. Xie, X.B. Jing, J.B. Buse, Z. Gu, H2O2-responsive vesicles integrated with transcutaneous patches for glucose-mediated insulin delivery, ACS Nano 11 (1) (2017) 613–620.
[30] Y.Q. Ye, J.C. Yu, C. Wang, N.Y. Nguyen, G.M. Walker, J.B. Buse, Z. Gu, Microneedles integrated with pancreatic cells and synthetic glucose-signal amplifiers for smart insulin delivery, Adv. Mater. 28 (16) (2016) 3115–3121.
[31] I.C. Lee, Y.C. Wu, S.W. Tsai, C.H. Chen, M.H. Wu, Fabrication of two-layer dissolving polyvinylpyrrolidone microneedles with different molecular weights for in vivo insulin transdermal delivery, RSC Adv. 7 (9) (2017) 5067–5075.
[32] S.X. Yang, F. Wu, J.G. Liu, G.R. Fan, W. Welsh, H. Zhu, T. Jin, Phase-transition microneedle patches for efficient and accurate transdermal delivery of insulin, Adv. Funct. Mater. 25 (29) (2015) 4633–4641.
[33] M.H. Ling, M.C. Chen, Dissolving polymer microneedle patches for rapid and efficient transdermal delivery of insulin to diabetic rats, Acta Biomater. 9 (11) (2013) 8952–8961.
[34] J.C. Yu, Y.Q. Zhang, W.J. Sun, A.R. Kahkoska, J.Q. Wang, J.B. Buse, Z. Gu, Insulin-responsive glucagon delivery for prevention of hypoglycemia, Small 13 (19) (2017).
[35] F.O. Obiweluozor, A. GhavamiNejad, S. Hashmi, M. Vatankhah-Varnoosfaderani, F.J. Stadler, A NIPAM-zwitterion copolymer: rheological interpretation of the specific ion effect on the LCST, Macromol. Chem. Phys. 215 (21) (2014) 2125. vol. 215, p. 1077, 2014.
[36] S. Hashmi, M. Vatankhah-Varnoosfaderani, A. GhavamiNejad, F.O. Obiweluozor, B. Du, F.J. Stadler, Self-associations and temperature dependence of aqueous solutions of zwitterionically modified N-isopropylacrylamide copolymers, Rheol. Acta 54 (6) (2015) 501–516.
[37] L. Du, A. Ghavaminejad, M. Vatankhah-Varnoosfaderani, F.J. Stadler, Study of the interactions of zwitterions and carbon nanotubes by nonlinear rheology in an aqueous environment, Langmuir 35 (5) (2019) 1964–1972.
[38] P. Zhang, F. Sun, C. Tsao, S.J. Liu, P. Jain, A. Sinclair, H.C. Hung, T. Bai, K. Wu, S.Y. Jiang, Zwitterionic gel encapsulation promotes protein stability, enhances pharmacokinetics, and reduces immunogenicity, Proc. Natl. Acad. Sci. U.S.A. 112 (39) (2015) 12046–12051.

[39] A.R. Unnithan, A.G. Nejad, A.R.K. Sasikala, R.G. Thomas, Y.Y. Jeong, P. Murugesan, S. Nasseri, D.M. Wu, C.H. Park, C.S. Kim, Electrospun zwitterionic nanofibers with in situ decelerated epithelialization property for non-adherent and easy removable wound dressing application, Chem. Eng. J. 287 (2016) 640–648.
[40] A. GhavamiNejad, J. Li, B. Lu, L. Zhou, L. Lam, A. Giacca, X.Y. Wu, Glucose-responsive composite microneedle patch for hypoglycemia-triggered delivery of native glucagon, Adv. Mater. (2019) e190151.
[41] A. GhavamiNejad, J. Li, B. Lu, A. Giacca, X.Y. Wu, Glucose regulation by modified boronic acid-sulfobetaine zwitterionic nanogels—a non-hormonal strategy for the potential treatment of hyperglycemia, Nanoscale 11 (2019) 10167–10171.
[42] C. Wang, Y.Q. Ye, G.M. Hochu, H. Sadeghifar, Z. Gu, Enhanced cancer immunotherapy by microneedle patch-assisted delivery of anti-PD1 antibody, Nano Lett. 16 (4) (2016) 2334–2340.
[43] M.C. Chen, Z.W. Lin, M.H. Ling, Near-infrared light-activatable microneedle system for treating superficial tumors by combination of chemotherapy and photothermal therapy, ACS Nano 10 (1) (2016) 93–101.
[44] Y. Hao, M.L. Dong, T.Y. Zhang, J.R. Peng, Y.P. Jia, Y.P. Cao, Z.Y. Qian, Novel approach of using near-infrared responsive PEGylated gold nanorod coated poly (L-lactide) microneedles to enhance the antitumor efficiency of docetaxel-loaded MPEG-PDLLA micelles for treating an A431 Tumor, ACS Appl. Mater. Interfaces 9 (18) (2017) 15317–15327.
[45] Y. Ye, C. Wang, X. Zhang, Q. Hu, Y. Zhang, Q. Liu, D. Wen, J. Milligan, A. Bellotti, L. Huang, G. Dotti, Z. Gu, A melanin-mediated cancer immunotherapy patch, Sci. Immunol. 2 (17) (2017).
[46] A. GhavamiNejad, M. SamariKhalaj, L.E. Aguilar, C.H. Park, C.S. Kim, pH/NIR light-controlled multidrug release via a mussel-inspired nanocomposite hydrogel for chemo-photothermal cancer therapy, Sci. Rep. 6 (2016).
[47] J.W. Lee, S.O. Choi, E.I. Felner, M.R. Prausnitz, Dissolving microneedle patch for transdermal delivery of human growth hormone, Small 7 (4) (2011) 531–539.

CHAPTER 4

Nanohybrid scaffold structures for smart drug delivery applications

S. Sowmya
Research Scientist, Amrita School of Dentistry, Amrita Vishwa Vidyapeetham, Kochi, India

Contents

4.1 Introduction

Scaffold is one of the integral parts of tissue engineering and regenerative medicine, as it can provide structural support at the defect site and the biological cues for regulating cell functions. Previous studies have extensively assessed the physiochemical and biological properties of scaffolds, which help in framing the prerequisite characteristics towards scaffold design. Wide varieties of biodegradable polymers, ceramics, or composites in combination with bioactive molecules are being investigated towards engineering scaffolds using various synthetic methods. The fact that properties such as porosity, surface roughness, and high surface area are not only found to influence cell-material interaction, but also play a key role in controlling the release of drug or other therapeutic agents augmented the application of scaffolds in medical field [1, 2]. The main advantage of scaffold-based delivery system is the possibility of site-specific delivery of drug at the affected site with high efficiency and minimal side effects. Moreover, besides serving as a surrogate extracellular matrix for tissue regeneration or repair, scaffold-based delivery system offer controlled release of drugs, thereby facilitating the management of diseases in a better way [3–5].

Drug release from the scaffold could be influenced by factors like porosity, degradation, drug-scaffold interaction, and the mode of drug

Biomimetic Nanoengineered Materials for Advanced Drug Delivery
https://doi.org/10.1016/B978-0-12-814944-7.00004-7

incorporation into the scaffold. Based on the pharmacokinetic and therapeutic requirements, drug release from the scaffold can be modulated. However, the introduction of stimulant-responsive scaffold systems has opened up excellent opportunities toward engineering smart delivery vehicles. In this chapter, we have discussed drug delivery from scaffolds under the influence of physiochemical properties, electrical stimulation, and magnetic stimulation.

4.2 Drug delivery under electrical stimulation

Drug release under the influence of electrical stimulation offers advantage on conditions which demand delivery of drug at the defect site when needed. The emergence of conducting scaffolds provides opportunity to modulate scaffold functionality through external electric field. Conducting polymers such as polypyrrole and polyaniline have already proven their potential in regulating cell-material interaction. Intrinsic electrical properties of bone and neuronal tissues and their ability to respond to external electrical field were exploited in the field of tissue engineering [6]. Redox properties of conducting scaffolds facilitate drug release as well [7]. Conducting polymers such as polypyrrole are being widely investigated as its redox and actuating properties are known to alter its electron transport properties, thereby controlling the drug release using applied electric field [8]. Electrically controlled drug release using polypyrrole was demonstrated by Sirivisoot et al. Antibiotics (penicillin/streptomycin) or an anti-inflammatory drug (dexamethasone) were immobilized within polypyrrole and electrodeposited on titanium. The results showed controlled release of both penicillin/streptomycin and dexamethasone upon electrical stimulation. Moreover, polypyrrole-dexamethasone enhanced the osteoblast adhesion and polypyrrole-penicillin/streptomycin inhibited fibroblast adhesion when compared to titanium [9]. The spontaneous redox property of polypyrrole upon electrical stimulation was also utilized for programmable and site-specific delivery of nerve growth factors. Growth factor-loaded mesoporous silica nanoparticles were incorporated with polypyrrole film and the composite showed enhanced release of growth factors upon electric stimulation and facilitated neurite outgrowth [10]. Drug release profile from graphene oxide (GO) nanosheets were improved by doping drug-loaded GO to polypyrrole matrix, wherein polypyrrole directed the electrical stimulation to trigger drug (dexamethasone) release whereas GO nanosheet acted as an efficient carrier for the drug. During nanocomposite preparation, positive charges grown on the polypyrrole chain are balanced by reaction with anionic dexamethasone (drug) and GO. The reduction of polypyrrole caused by electrical stimulation neutralizes the charges and triggered the release of the drug, Dexamethasone [11].

Servant et al. have utilized graphene for developing electro-active hydrogel scaffold, owing to its high electron mobility. Pristine graphene was incorporated during in situ polymerization to form methacrylic acid hydrogel. The hybrid gel containing 0.2 mg/mL of graphene showed significant enhancement in the release of ^{14}C-sucrose (model drug molecule) when an external electric field of 10 V was applied for 4 min. The release of amphiphilic drug, radiolabeled ^{14}C-doxorubicin, triggered by electrical stimulation explained the ability of graphene hybrid hydrogel system to modulate drug release with different hydrophobic characteristics. The study also explained that initial burst release of drug molecules from the system was due to electro-responsive character of the hydrogel rather than the structural damage. More importantly, the excellent heat dissipation property of graphene helped to circumvent the possible temperature rise during electrical stimulation and demonstrated on-demand drug release from the hydrogel without damaging the host tissue [12].

4.3 Drug delivery under magnetic stimulation

Polymer gels embedded with magnetic nanoparticles, also termed as ferrogels, are considered as excellent tool for drug delivery under the influence of magnetic stimulation. One of the key mechanisms of drug release from ferrogels is the deformation and volumetric change occurring to the gel when external magnetic field is applied [13, 14]. Water convection through the pores, resulting from the deformation, aids in the release of drug or other biomolecular agents from the gel, that is, when there is no magnetic field the drug release profile from the gel will display normal mode of diffusion, since the magnetic moments in the ferrogel will be randomly oriented. But, when the magnetic field is turned on, the magnetic moments of particles align with the magnetic field and cause the aggregation of particles within the gel, which in turn results in the decrease in porosity of the gel. This reduction in pore size of the gel will restrict the release of drug because the drug particle will get confined within the polymeric network. Subsequently, when the magnetic field is turned off, the pores will reopen and instant release of drug occurs [15, 16]. Also, it was found that the drug release profile from the gel is greatly influenced by the switch duration time (SDT), which is defined as the time duration between each on-off cycle. Triggering drug delivery using a noncontact, switch-on-switch-off mode adds on to the advantage of ferrogels for controlled drug delivery using external magnetic field. Previous studies have reported the utilization of a variety of natural and synthetic polymers toward the development of different types of ferrogels and showed superior controlled release of drugs when subjected to magnetic field. Besides static magnetic field, the use of alternating magnetic field (AMF) provides additional opportunities to actuate magnetic nanoparticles, thereby modulating the drug release profile from hybrid gels. The application of high-frequency AMF not only facilitates the penetration deep into the body without affecting normal tissue, but also was utilized to control drug release behavior of the magnetic nanoparticle-loaded gels based on thermo-responsive polymers such as poly(*N*-isopropylacrylamide). Gel collapse and subsequent drug release occurs when heat generation due to applied AMF goes beyond lower critical transition temperature of the polymer. Temperature response of the gel to get desired drug release profile was demonstrated by tailoring the composition of polymer and/or loading of magnetic nanoparticles into it [17]. Studies have also focused on improving cell adhesion property and transport of large molecules from the gels. The conjugation of RGD (Arginine-Glycine-Aspartic) peptides or other bioactive molecules with polymer matrix and the development of macroporous gels are known to enhance the cell adhesion properties and improve release kinetics of drugs, respectively, from the gel. Ferrogels with porous structures are reported to have improved magnetic sensitivity, as the nanoparticles residing at the pore walls in higher concentration facilitate enhanced deformation and fluid flow [15]. For the most part, previous studies have described magnetically controlled drug release profile of hydrophilic drugs from polymeric gels and investigated the influence of various physiochemical parameters to improve the efficacy of the system. Qin et al. have addressed the challenge of incorporating hydrophobic drug into ferrogel by employing pluronic as gelling material. They utilized the ability of pluronic copolymers to aggregate into micelles at above certain concentration and temperature. The water-free core of the aggregates is known to solubilize hydrophobic drugs. The study reported the development of a super paramagnetic iron oxide loaded pluronic gel and controlled release of drug, indomethacin under the influence of applied magnetic field [18].

4.4 pH-sensitive scaffolds for drug delivery

pH-dependent drug delivery systems offer improved therapeutic efficiency as they facilitate site-specific targeting and drug release kinetics in response to pathophysiological factors. The triggering factor for drug release is the variation in pH in the body fluid or segments in tissues. Polymers having acidic (carboxylic/sulfonic) or basic (ammonium salts) functional groups in the backbone have been utilized for pH-dependant drug delivery as the alteration in pH causes swelling or de-swelling of these polymers thereby aiding drug release [19]. Delivery systems having acidic functional group will swell with increasing pH and the other way round. Apart from investigating derivatives of polyacrylic acid and cellulose with copolymers, pH-sensitive drug carriers were reported using chemical modification or functionalization of composite systems. Moreover, nanofibrous scaffolds developed by incorporating sodium bicarbonate into polymers resulted in acid-responsive system which showed potential targeted drug delivery at tumor microenvironment, inflammatory disease, and other pathophysiological conditions where intracellular pH is lower than normal physiological pH. At low pH, sodium bicarbonate reacts with protons and release carbon dioxide. This in turn may rupture the gel network and porosity of the carrier and trigger drug release [20]. Core-shell electrospun system also described dual delivery of anti-inflammatory and antitumor drugs in response to alteration in pH [21]. Gelatin fiber loaded with anti-inflammatory drug along with sodium biocarbonate was the shell and the antitumor drug incorporated with polycaprolactone fiber was the core of the system. Acidic pH in the tumor environment facilitated collapse of shell layer, thereby releasing anti-inflammatory drug, which might help to control the inflammatory response following tumor resection. Subsequently, sustained release of antitumor drug happened from the core [21]. Drug release from the scaffold in response to pH change will follow diffusion controlled release, if scaffold pore size is higher than the molecular dimension of the drug. Drug release can occur through swelling mechanism as well [22, 23]. Enzyme-mediated pH modulation for eroding the pH-sensitive polymer is also being utilized for controlled drug delivery. The pH change, resulting from a substrate-enzyme reaction, can erode and degrade the polymer network, thereby triggering the release of dispersed therapeutic agent from the gel [24]. Intracellular drug delivery system has advanced with the integration of drug-loaded nanoparticles with pH-sensitive scaffolds. These systems have utilized the pH difference between intra- and extracellular environments for triggering drug release by protonation, acid labile bond cleavage, and dissociation of delivery vehicle [25].

References

[1] T. Garg, O. Singh, S. Arora, R. Murthy, Scaffold: a novel carrier for cell and drug delivery, Crit. Rev. Ther. Drug Carrier Syst. 29 (1) (2012) 1–63.

[2] J.D. Kretlow, L. Kloud, A.G. Mikos, Injectable matrices and scaffolds for drug delivery in tissue engineering, Adv. Drug Deliv. Rev. 59 (4–5) (2007) 263–273.

[3] M. Goldberg, R. Langer, X. Jia, Nanostructured materials for applications in drug delivery and tissue engineering, J. Biomater. Sci. 18 (3) (2007). Polymer Edition.

[4] V. Mouriño, A.R. Boccaccini, Bone tissue engineering therapeutics: controlled drug delivery in three-dimensional scaffolds, J. R. Soc. Interface 7 (43) (2010) 209–227.

[5] J.S. Boateng, K.H. Matthews, H.N.E. Stevens, G.M. Eccleston, Wound healing dressings and drug delivery systems: a review, J. Pharm. Sci. 97 (8) (2008) 2892–2923.

[6] N.K. Guimard, N. Gomez, C.E. Schmidt, Conducting polymers in biomedical engineering, Prog. Polym. Sci. 32 (2007) 876–921.

[7] R. Wadhwa, C.F. Lagenaur, X.T. Cui, Electrochemically controlled release of dexamethasone from conducting polymer polypyrrole coated electrode, J. Control. Release 110 (2006) 531–541.

[8] D.D. Ateh, A. NavsariaH, P. Vadgama, Polypyrrole-based conducting polymers and interactions with biological tissues, J. R. Soc. Interface 3 (2006) 741–752.

[9] S. Sirivisoot, R. Pareta, T.J. Webster, Electrically controlled drug release from nanostructured polypyrrole coated on titanium, Nanotechnology 22 (2011) 085101 (15 p).

[10] Y. Cho, R. Shi, A. Ivanisevic, R.B. Borgens, A mesoporous silica nanosphere-based drug delivery system using an electrically conducting polymer, Nanotechnology 20 (2009) 275102 (12 p).

[11] C.L. Weaver, L.R. JM, X. Luo, X.T. Cui, Electrically controlled drug delivery from graphene oxide nanocomposite films, ACS Nano 8 (2) (2014) 1834–1843.

[12] A. Servant, V. Leon, D. Jasim, L. Methven, P. Limousin, E.V. Fernandez-Pacheco, M. Prato, K. Kostarelos, Graphene-based electroresponsive scaffolds as polymeric implants for on-demand drug delivery, Adv. Healthc. Mater. 3 (2014) 1334–1343.

[13] N.S. Satarkar, J.Z. Hilt, Magnetic hydrogel nanocomposites for remote controlled pulsatile drug release, J. Control. Release 130 (3) (2008) 246–251.

[14] S.-H. Hu, T.-Y. Liu, D.-M. Liu, S.-Y. Chen, Controlled pulsatile drug release from a ferrogel by a high-frequency magnetic field, Macromolecules 40 (19) (2007) 6786–6788.

[15] S.-H. Hu, T.-Y. Liu, D.-M. Liu, S.-Y. Chen, Nano-ferrosponges for controlled drug release, J. Control. Release 121 (2007) 181–189.

[16] X. Zhaob, J. Kim, C.A. Cezar, N. Huebsch, K. Lee, K. Bouhadir, D.J. Mooney, Active scaffolds for on-demand drug and cell delivery, PNAS 108 (1) (2011) 67–72.

[17] M. Uva, D. Pasqui, L. Mencuccini, S. Fedi, R. Barbucci, Influence of alternating and static magnetic fields on drug release from hybrid hydrogels containing magnetic nanoparticles, J. Biomater. Nanobiotechnol. 5 (2014) 116–127.

[18] J. Qin, I. Asempah, S. Laurent, A. Fornara, R.N. Muller, M. Muhammed, Injectable superparamagnetic ferrogels for controlled release of hydrophobic drugs, Adv. Mater. 21 (2009) 1354–1357.

[19] B. Veeranna, M.P. Venkatesh, N. Vishal Gupta, K.L. Krishna, pH sensitive drug delivery systems: a review, Am. J. Drug Discovery Dev. 1 (1) (2011) 24–48.

[20] Q. Sang, G.R. Williams, H. Wu, et al., Electrospun gelatin/sodium bicarbonate and poly(lactide-co-e-caprolactone)/sodium bicarbonate nanofibers as drug delivery systems, Mater. Sci. Eng. C Mater. Biol. Appl. 81 (2017) 359–365.
[21] Q. Sang, H. Li, G. Williams, H. Wu, L.-M. Zhu, Core-shell poly(lactide-coe-caprolactone)-gelatin fiber scaffolds as pH-sensitive drug delivery systems, J. Biomater. Appl. 32 (8) (2018) 1105–1118.
[22] A. Gutowska, Y.H. Bae, J. Feijen, S.W. Kim, Heparin release from thermosensitive hydrogels, J. Control. Release 22 (1992) 95–104.
[23] L. Ferreira, M.M. Vidal, M.H. Gil, Evaluation of poly(2-hydroxyethyl methacrylate) gels as drug delivery systems at different pH values, Int. J. Pharm. 194 (2000) 169–180.
[24] P.S. Pramod, N.U. Deshpande, M. Jayakannan, Real-time drug release analysis of enzyme and ph responsive polysaccharide nanovesicles, J. Phys. Chem. B 119 (33) (2015) 10511–10523.
[25] M. Zhang, J. Liu, Y. Kuang, Q. Li, D.-W. Zheng, Q. Song, H. Chen, X. Chen, Y. Xu, C. Li, B. Jiang, Ingenious pH-sensitive dextran/mesoporous silica nanoparticles based drug delivery systems for controlled intercellular drug release, Int. J. Biol. Macromol. 98 (2017) 691–700.

CHAPTER 5

3D bioprinting for active drug delivery

Joshua Lee[a], Afeesh Rajan Unnithan[a,b,c], Chan Hee Park[a,b], Cheol Sang Kim[a,b]
[a]Department of Bionanosystem Engineering, Graduate School, Chonbuk National University, Jeonju, Republic of Korea
[b]Division of Mechanical Design Engineering, Chonbuk National University, Jeonju, Republic of Korea
[c]Healthcare Technology Institute, Department of Chemical Engineering, University of Birmingham, Birmingham, United Kingdom

Contents

5.1 The advent of three-dimensional bioprinting

The field of three-dimensional (3D) bioprinting, also known simply as bioprinting, became a hot topic once scientists envisioned using the technology to create artificial functional tissues and organs, and forever changed the tissue engineering field and the biofabrication landscape. Fabrication of functional tissues and organs from biomaterials is a controversial topic that comes with unprecedented ethical challenges which has its opposers as well as supporters. Nevertheless, the technology still has a long way to go before the realization of the transplantation of fully functional tissues.

Another use of 3D bioprinting is the creation of drug-releasing structures or scaffolds for the treatment of injuries or regenerative medicine. It is now being realized how much 3D bioprinting can offer to the field of pharmaceuticals and drug discovery.

Biomimetic Nanoengineered Materials for Advanced Drug Delivery
https://doi.org/10.1016/B978-0-12-814944-7.00005-9

5.1.1 Additive manufacturing

Although additive manufacturing (AM) is now synonymous with 3D printing (3DP), the term was initially used to refer to any process that involves "adding" layers upon layers of materials to create a new 3D object. However, once 3DP became a household word and the primary AM technology, the two became synonymous in use. Here, we will also talk about other AM technologies such as stereolithography (STL) and laser sintering, and how they paved the way for 3D bioprinting.

5.1.1.1 Stereolithography

While traditional 3DP consists of laying down layers of polymer materials, STL is slightly different in that it uses photopolymerization to cross-link polymers into a 3D solid. An ultraviolet laser is focused on a vat of photopolymer resin, and with the help of computer-aided design (CAD), the laser draws the shape layer by layer. When one layer is done, the platform lowers and is coated with another layer of resin by a blade so that the process can be repeated to build the 3D shape. Although this technology is generally used to create 3D structures of plastics or resins, STL can be used with soft materials such as hydrogels for applications in tissue engineering [1–3]. This technology can also be used with drug-loaded hydrogels to create structures that will release the drug as the hydrogel in which it is encapsulated degrades or dissolves over time, as shown later in the chapter.

5.1.1.2 Fused deposition modeling

Fused deposition modeling (FDM) is a 3DP process that utilizes filament-type thermoplastics with a printer head that is capable of heating the filament to melting temperatures to create a structure from the bottom up. The printer nozzle moves back and forth according to the design described in the STL file format, adding layer by layer the filament until the structure is complete. The FDM is a precursor to the most widely used bioprinting technique, extrusion-based printing. The technologies are very similar except that extrusion printing does not usually need a heating element as most bioinks will not need to be melted for printing.

5.1.1.3 3D bioprinting

As AM technology and the materials science and biology fields advanced, biofabrication of complex biological products has become the new hot topic in the field of tissue engineering [4]. Bioprinting works very much like FDM, but uses materials called bioinks instead of thermoplastics to create artificial tissue analogs used for regenerative medicine and drug screening among other applications [5–7].

The most popular use of bioprinting currently is to create highly biocompatible scaffolds and tissue replacements. There is a great need for tissues and organs that are suitable for transplantation and bioprinting is a technology that shows great promise in fulfilling that need [8], although it may take more than a decade to do so. In order for this to happen, advances in biomaterials, cell biology, and bioprinter technology are necessary.

To create a bioprinted construct that is suitable and transplantable in a human body, the construct should be made of biomimetic materials. Focus on creating constructs that mimic the natural extracellular matrix (ECM) has only increased with new bioprinting technology. Materials with ECM-like characteristics have been long sought after and some researchers have even utilized decellularized ECM (dECM) in pursuit of creating the best environment for cell growth [9–11].

5.2 Current bioprinting research

5.2.1 Materials

One of the most popular materials used to create bioinks is hydrogel, based on both natural and synthetic polymers. The most well-documented polymers for use as bioinks in recent years are gelatin [12, 13], alginate [14–16], and hyaluronic acid (HA) [17, 18] because of their biocompatibility, availability, and cost-effectiveness. They have also been modified to possess unique properties such as different cross-linking mechanisms via methods such as methacrylation. Such a hydrogel is tunable, depending on how many of these functional groups are conjugated, by varying the reaction time or amount of the reactant. One example of this is gelatin methacryloyl, or better known as GelMA [7, 19, 20], which could be reversibly cross-linked via temperature and permanently cross-linked by UV (ultraviolet) irradiation.

One of the greatest advantages these dual-mechanism materials provide is the printing of stable structures without special coaxial nozzles or the addition of catalysts/initiators. GelMA is reversibly cross-linked at a temperature of around 22–27°C, so one can slightly heat the solution before printing so that it flows, but solidifies as it comes out of the nozzle at a lower temperature. Then as the structure begins to take shape, it can be cross-linked again by UV irradiation to provide more secure structural integrity.

Alginate, or alginic acid, is another polymer that can be methacrylated [21]. Once methacrylated, it can be cross-linked in two ways: via ionic interactions and UV irradiation. Ionic cross-linking can be achieved through the addition of divalent cations such as calcium and barium. Many researchers have used this effect to create 3D structures that are generally biocompatible and have low toxicity, as shown in Fig. 5.1 [14, 22]. A popular method of printing alginate hydrogels utilizes CaCl bath for instant cross-linking upon printing for structural integrity, while another widely used method uses coaxial nozzles for the co-printing of the CaCl solution and alginate hydrogel.

Another material used to make hydrogels is HA, which also has been methacrylated for use in printing applications [23]. HA is a natural glycosaminoglycan that can be found in almost all connective tissues and is one of the chief components of the ECM, playing a role in regulating many cellular behaviors and tissue functions. These advantages make them great candidates as biocompatible hydrogels, but their viscoelastic properties limit their use as bioink in 3D bioprinting.

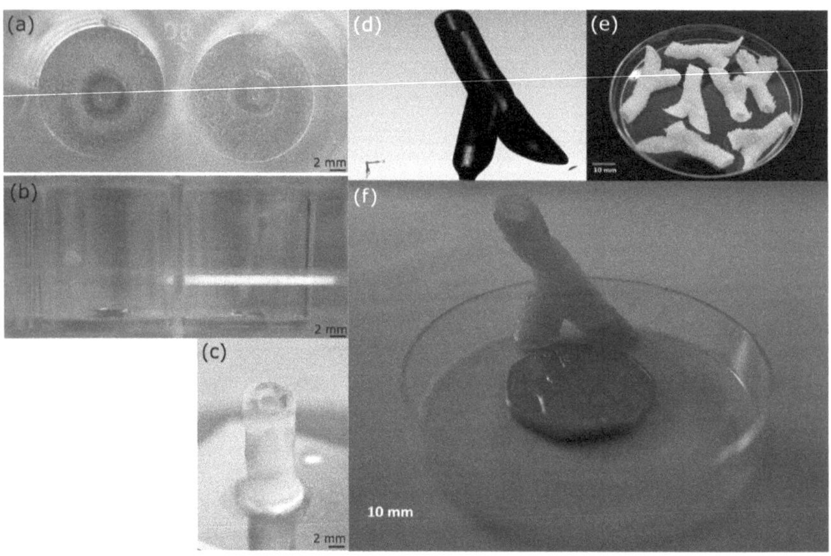

Fig. 5.1 (A–C) Printed alginate structures cross-linked via calcium chloride. (D) Computer-aided design image of a tube structure for printing. (E and F) Printed vascular structures. *(Adapted with permission from A.G. Tabriz, M.A. Hermida, N.R. Leslie, W.M. Shu, Three-dimensional bioprinting of complex cell laden alginate hydrogel structures. Biofabrication 7 (4) (2015). 045012 and A. Faulkner-Jones, C. Fyfe, D.J. Cornelissen, J. Gardner, J. King, A. Courtney, W.M. Shu, Bioprinting of human pluripotent stem cells and their directed differentiation into hepatocyte-like cells for the generation of mini-livers in 3D. Biofabrication 7 (4) (2015) 044102, licensed under CC BY 3.0).*

What better way to mimicking the ECM for accelerated and optimal cell growth than to use natural ECM itself? Many researchers have turned to dECM as a potential material for bioink use [9, 24]. The dECM materials are produced by first removing the cellular material of the ECM material while minimizing ECM loss and damage. Some of the dECM used in the literature come from adipose, heart, and cartilage tissues using a combination of detergents, shaking and rinsing to decellularize the tissues. The advantages of using dECM include providing an environment that is familiar to cells and being able to bind growth factors and slowly release them to the cells over time.

5.2.2 Applications

5.2.2.1 Effects on cells

Since the act of bioprinting exerts a shear force on the material inside, there have been studies detailing and explaining the effects on the cells that are printed within the bioink. One such effect is the alignment of printed cells during the printing process. Due to the shear force enacted upon the bioink and the cells encapsulated, the printed cells are aligned, with or without the help of external factors such as grooves [25–27]. Some aligned cells, such as nerve and heart cells, have been shown to show significantly more differentiation than their randomly oriented counterparts. If printing can offer cell alignment when needed, it can be a great asset to the user. 3D bioprinted differentiated C2C12 myotubes shown in Fig. 5.2 were described as ideal for artificial skeletal muscles [28]. In a separate study, C2C12 cells were shown to align as the cells were exposed to shear stresses during the printing process, allowing them to acquire a highly elongated shape, as depicted in Fig. 5.3 [26].

Another study looked at the bioprinting of growth factors using an inkjet bioprinter onto patterned microfibers to direct stem cell fate via geometric and biochemical cues [27]. Fibers were arranged so that myocytes would align themselves since they need to generate force in a specific direction and tenocytes and osteoblasts could build highly oriented collagen matrices to withstand the mechanical load demanded by the body. With the addition of specific bioprinted growth factors, the cells were further encouraged to differentiate into their respective specialized cells.

5.2.2.2 Tissue engineering

The greatest potential of 3D printing can be realized possibly in the field of tissue engineering, which is defined as an interdisciplinary field that brings together engineering and the life sciences to develop biological substitutes

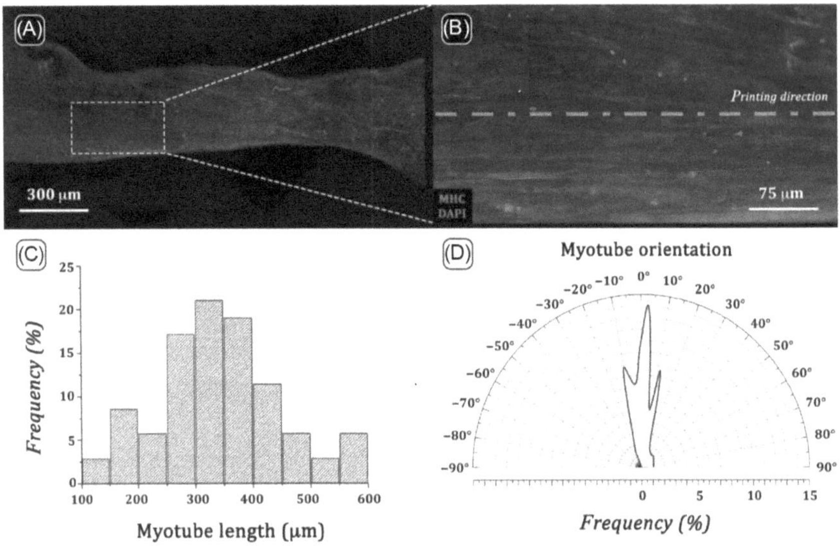

Fig. 5.2 (A) Tiled immunofluorescence micrographs of a large region (300 × 2000 μm) and (B) high magnification of the ROI (region of interest) evidenced in part (A). (C) Myotube length distribution. (D) Polar chart of myotube orientation, with 0 degree corresponding to the printing direction. *(Reproduced with permission from M. Costantini, S. Testa, P. Mozetic, A. Barbetta, C. Fuoco, E. Fornetti, F. Tamiro, S. Bernardini, J. Jaroszewicz, W. Swieszkowski, M. Trombetta, L. Castagnoli, D. Seliktar, P. Garstecki, G. Cesareni, S. Cannata, A. Rainer, C. Gargioli, Microfluidic-enhanced 3D bioprinting of aligned myoblast-laden hydrogels leads to functionally organized myofibers in vitro and in vivo. Biomaterials 131 (2017) 98–110, licensed under CC BY-NC-ND 4.0).*

that restore, maintain, or improve tissue function [29]. Many scaffolds mimicking various tissues and organs such as cartilage [30] and skin [31] have been fabricated.

5.3 3D bioprinting for active drug delivery

5.3.1 Current research

While 3D printing and bioprinting is somewhat advanced in the field of tissue engineering, it is not yet the go-to method for creating drug delivery devices. One group that has utilized this technology has printed intrauterine devices (IUDs) loaded with estrogen and progesterone in different sizes and designs that can be tailored to the patient [32]. In this particular case, PCL pellets were loaded with the hormones using a coating of silicone oil and vortexing to stick the drug powder onto the pellets. Then, the coated pellets are melt extruded and printed with varying amounts of dosages and different

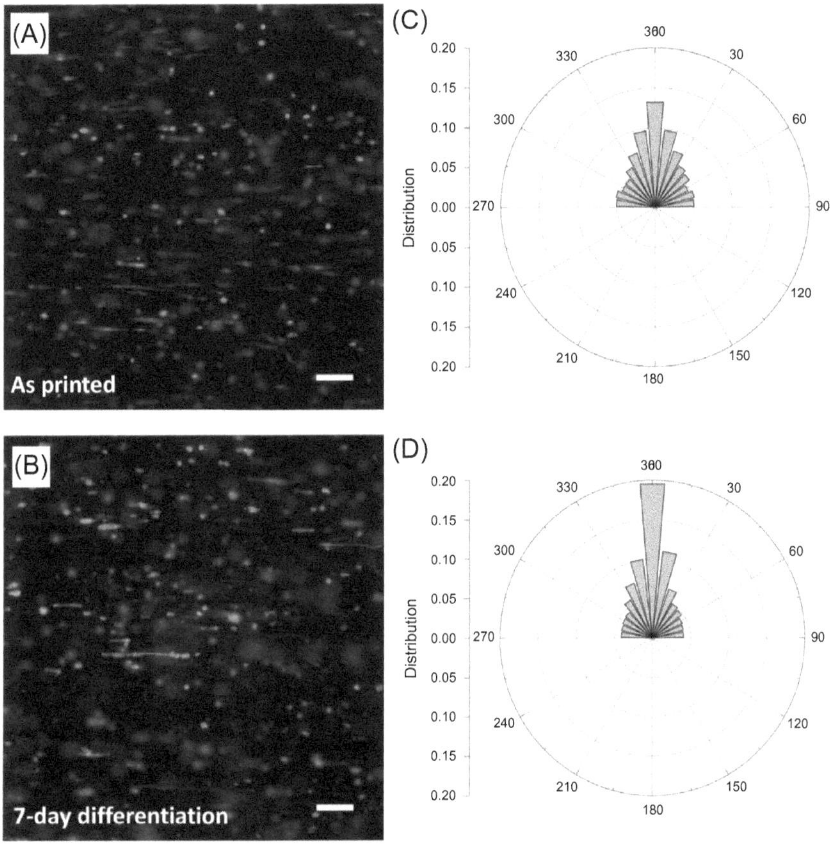

Fig. 5.3 (A and B) Fluorescence images of printed and differentiated C2C12 myoblast cells, respectively. (C and D) Rose plots of the myoblast orientation of (A) and (B), respectively. *(Reproduced with permission from P. Mozetic, S.M. Giannitelli, M. Gori, M. Trombetta, A. Rainer, Engineering muscle cell alignment through 3D bioprinting. J. Biomed. Mater. Res. A 105 (9) (2017) 2582–2588, Copyright Wiley Periodicals 2017).*

shapes as shown in Fig. 5.4. Three different hormones were used: estrone (E1), estradiol (E2), and estriol (E3).

Yet another group used a different kind of 3D printing to create drug-loaded hydrogels for the delivery of ibuprofen. Martinez et al. utilized STL 3D printing technology to fabricate hydrogels containing up to 30% w/w water and 10% w/w ibuprofen using polyethylene glycol diacrylate (PEGDA) and polyethylene glycol (PEG300) with riboflavin and triethanolamine (TEA) as photoinitiators [33]. Their goal was to create alternatives to tablets and two-piece capsules for oral delivery of drugs, which are inflexible in their dosage and limited to only a few dosage strengths. With their

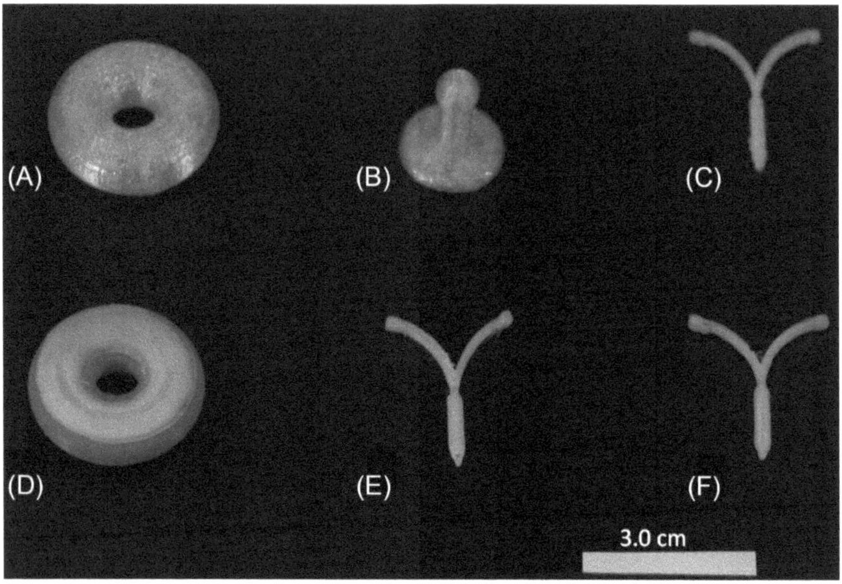

Fig. 5.4 IUDs in different shapes and with or without various drugs. (A) Control donut-shaped pessary, (B) Gellhorn-shaped pessary, (C) IUD, (D) pessary printed with a combination of filaments (*red; light gray* in print version—PLA; *white*—PCL-E2), (E) PCL-E1, and (F) PCL-E2. *(Reproduced with permission from K. Tappa, U. Jammalamadaka, D.H. Ballard, T. Bruno, M.R. Israel, H. Vemula, J.M. Meacham, D.K. Mills, P.K. Woodard, J.A. Weisman, Medication eluting devices for the field of OBGYN (MEDOBGYN): 3D printed biodegradable hormone eluting constructs, a proof of concept study. PLoS One 12 (8) (2017) e0182929, licensed under CC BY 4.0).*

drug-loaded hydrogels, the drug release profiles can be fine-tuned by altering the degree of cross-linking while using a highly convenient method of fabrication. They have succeeded in reproducing their results with different amounts of water content in the hydrogel, as shown in Fig. 5.5A.

Genina et al. have recently designed and printed a dual-compartmental dosage unit (dcDU) containing two different drugs: rifampicin (RIF) and isoniazid (ISO) [34]. The combination therapy of these two drugs ultimately improved the treatment efficacy since RIF is absorbed well from the stomach, while ISO is mostly absorbed while in the intestine, leading to a prolonged treatment period with one dcDU. The dcDU design consists of two separate compartments of water-insoluble PLA, capped off with a water-soluble PVA cap for the release of the drugs, as shown in Fig. 5.5B–D.

The 3D printing technology is also being used to deliver growth factors such as bone morphogenetic protein-2 (BMP-2) [35] and vascular endothelial growth factors (VEGFs) [36]. Bioprinting is extensively used because it

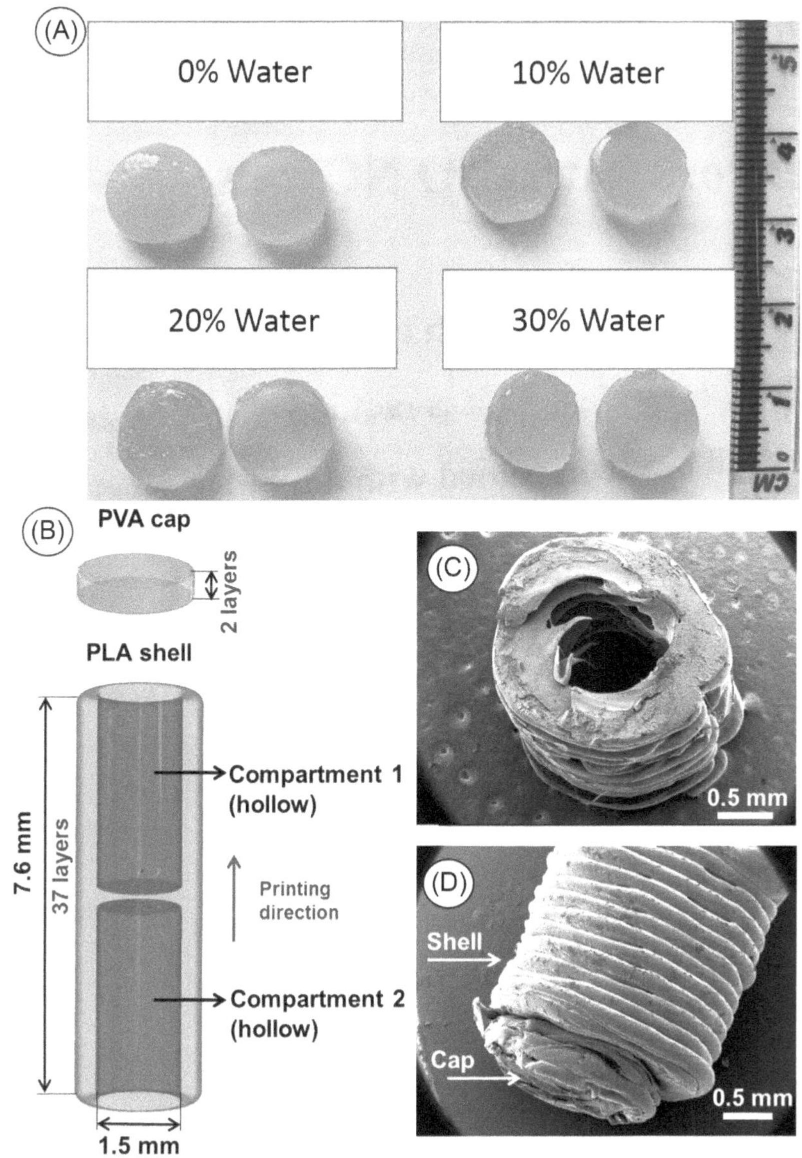

Fig. 5.5 (A) Hydrogels printed with varying amounts of water content with riboflavin/triethanolamine photoinitiator. (B) Schematic representation of the dual-compartmental dosage unit. (C and D) Scanning electron micrographs of the top and side views, respectively. *(Reproduced and adapted with permission from P.R. Martinez, A. Goyanes, A.W. Basit, S. Gaisford, Fabrication of drug-loaded hydrogels with stereolithographic 3D printing. Int. J. Pharm. 532 (1) (2017) 313–317 and N. Genina, J.P. Boetker, S. Colombo, N. Harmankaya, J. Rantanen, A. Bohr, Anti-tuberculosis drug combination for controlled oral delivery using 3D printed compartmental dosage forms: from drug product design to in vivo testing. J. Control. Release 268 (2017) 40–48, Copyright Elsevier 2017).*

primarily uses a hydrogel or a combination of hydrogels as bioinks, which can hold great amounts of drug/growth factor and release them at a relatively slow rate, compared to conventional methods all the while releasing them at the site the drug is needed most. The ability to create patient-tailored scaffolds and drug-loaded units is another reason that 3D bioprinting is up-and-coming in the field of pharmaceuticals and drug delivery.

5.3.2 Future outlook

In the technology-heavy world we live in today, everything we have and use is personalized and tailored to us. The advertisements we see online are based on our site surfing history, the apps we use on our phone know our name and address, and our phones know our habits better than we do ourselves. With 3D bioprinting technology, we can now have tissue analogs and drugs tailored to our needs. Not every person requires the same amount of drug, just as we need a different amount of food in our diet.

Although the technology has been around for over 10 years, it is still in its infancy and adoption of the tech is still considerably low. Research into how to use 3D printing to its full potential is still required, but the future looks to be bright, with many research groups around the world studying, experimenting, and producing promising results that could see real-world use.

References

[1] B. Dhariwala, E. Hunt, T. Boland, Rapid prototyping of tissue-engineering constructs, using photopolymerizable hydrogels and stereolithography, Tissue Eng. 10 (9-10) (2004) 1316–1322.

[2] R.J. Mondschein, A. Kanitkar, C.B. Williams, S.S. Verbridge, T.E. Long, Polymer structure-property requirements for stereolithographic 3D printing of soft tissue engineering scaffolds, Biomaterials 140 (2017) 170–188.

[3] R. Gauvin, Y.C. Chen, J.W. Lee, P. Soman, P. Zorlutuna, J.W. Nichol, H. Bae, S.C. Chen, A. Khademhosseini, Microfabrication of complex porous tissue engineering scaffolds using 3D projection stereolithography, Biomaterials 33 (15) (2012) 3824–3834.

[4] V. Mironov, T. Trusk, V. Kasyanov, S. Little, R. Swaja, R. Markwald, Biofabrication: a 21st century manufacturing paradigm, Biofabrication 1 (2) (2009).

[5] L.E. Bertassoni, J.C. Cardoso, V. Manoharan, A.L. Cristino, N.S. Bhise, W.A. Araujo, P. Zorlutuna, N.E. Vrana, A.M. Ghaemmaghami, M.R. Dokmeci, A. Khademhosseini, Direct-write bioprinting of cell-laden methacrylated gelatin hydrogels, Biofabrication 6 (2) (2014).

[6] Y.S. Zhang, F. Davoudi, P. Walch, A. Manbachi, X. Luo, V. Dell'Erba, A.K. Miri, H. Albadawi, A. Arneri, X.Y. Li, X.Y. Wang, M.R. Dokmeci, A. Khademhosseini, R. Oklu, Bioprinted thrombosis-on-a-chip, Lab Chip 16 (21) (2016) 4097–4105.

[7] W.T. Jia, P.S. Gungor-Ozkerim, Y.S. Zhang, K. Yue, K. Zhu, W.J. Liu, Q. Pi, B. Byambaa, M.R. Dokmeci, S.R. Shin, A. Khademhosseini, Direct 3D bioprinting of perfusable vascular constructs using a blend bioink, Biomaterials 106 (2016) 58–68.
[8] H.W. Kang, S.J. Lee, I.K. Ko, C. Kengla, J.J. Yoo, A. Atala, A 3D bioprinting system to produce human-scale tissue constructs with structural integrity, Nat. Biotechnol. 34 (3) (2016) 312.
[9] F. Pati, J. Jang, D.H. Ha, S.W. Kim, J.W. Rhie, J.H. Shim, D.H. Kim, D.W. Cho, Printing three-dimensional tissue analogues with decellularized extracellular matrix bioink, Nat. Commun. 5 (2014).
[10] N. Rotter, R. Riepl, A. Elsasser, L. Koerber, R. Breiter, P. Krumm, A. Von Bomhard, S. Schwarz, Reconstruction of nasal septum cartilage using decellularised extracellular cartilage matrix (DECM) in a rabbit model, J. Tissue Eng. Regen. Med. 8 (2014) 496–497.
[11] E.C. Beck, M. Barragan, M.H. Tadros, S.H. Gehrke, M.S. Detamore, Approaching the compressive modulus of articular cartilage with a decellularized cartilage-based hydrogel, Acta Biomater. 38 (2016) 94–105.
[12] W. Schuurman, P.A. Levett, M.W. Pot, P.R. van Weeren, W.J.A. Dhert, D.W. Hutmacher, F.P.W. Melchels, T.J. Klein, J. Malda, Gelatin-methacrylamide hydrogels as potential biomaterials for fabrication of tissue-engineered cartilage constructs, Macromol. Biosci. 13 (5) (2013) 551–561.
[13] J.W. Nichol, S.T. Koshy, H. Bae, C.M. Hwang, S. Yamanlar, A. Khademhosseini, Cell-laden microengineered gelatin methacrylate hydrogels, Biomaterials 31 (21) (2010) 5536–5544.
[14] A.G. Tabriz, M.A. Hermida, N.R. Leslie, W.M. Shu, Three-dimensional bioprinting of complex cell laden alginate hydrogel structures, Biofabrication 7 (4) (2015) 045012.
[15] N.E. Fedorovich, J.R. Dewijn, A.J. Verbout, J. Alblas, W.J.A. Dhert, Three-dimensional fiber deposition of cell-laden, viable, patterned constructs for bone tissue printing, Tissue Eng. Part A 14 (1) (2008) 127–133.
[16] J. Jia, D.J. Richards, S. Pollard, Y. Tan, J. Rodriguez, R.P. Visconti, T.C. Trusk, M.J. Yost, H. Yao, R.R. Markwald, Y. Mei, Engineering alginate as bioink for bioprinting, Acta Biomater. 10 (10) (2014) 4323–4331.
[17] A. Skardal, J.X. Zhang, G.D. Prestwich, Bioprinting vessel-like constructs using hyaluronan hydrogels crosslinked with tetrahedral polyethylene glycol tetracrylates, Biomaterials 31 (24) (2010) 6173–6181.
[18] L. Pescosolido, W. Schuurman, J. Malda, P. Matricardi, F. Alhaique, T. Coviello, P.R. van Weeren, W.J.A. Dhert, W.E. Hennink, T. Vermonden, Hyaluronic acid and dextran-based semi-IPN hydrogels as biomaterials for bioprinting, Biomacromolecules 12 (5) (2011) 1831–1838.
[19] K. Yue, G. Trujillo-de Santiago, M.M. Alvarez, A. Tamayol, N. Annabi, A. Khademhosseini, Synthesis, properties, and biomedical applications of gelatin methacryloyl (GelMA) hydrogels, Biomaterials 73 (2015) 254–271.
[20] J. Kim, R. Gauvin, H.J. Yoon, J.H. Kim, S.M. Kwon, H.J. Park, S.H. Baek, J.M. Cha, H. Bae, Skin penetration-inducing gelatin methacryloyl nanogels for transdermal macromolecule delivery, Macromol. Res. 24 (12) (2016) 1115–1125.
[21] O. Jeon, K.H. Bouhadir, J.M. Mansour, E. Alsberg, Photocrosslinked alginate hydrogels with tunable biodegradation rates and mechanical properties, Biomaterials 30 (14) (2009) 2724–2734.
[22] Z.J. Wu, X. Su, Y.Y. Xu, B. Kong, W. Sun, S.L. Mi, Bioprinting three-dimensional cell-laden tissue constructs with controllable degradation, Sci. Rep. 6 (2016).
[23] L.L. Ouyang, C.B. Highley, C.B. Rodell, W. Sun, J.A. Burdick, 3D printing of shear-thinning hyaluronic acid hydrogels with secondary cross-linking, ACS Biomater. Sci. Eng. 2 (10) (2016) 1743–1751.

[24] A. Skardal, M. Devarasetty, H.W. Kang, I. Mead, C. Bishop, T. Shupe, S.J. Lee, J. Jackson, J. Yoo, S. Soker, A. Atala, A hydrogel bioink toolkit for mimicking native tissue biochemical and mechanical properties in bioprinted tissue constructs, Acta Biomater. 25 (2015) 24–34.
[25] R. Bhuthalingam, P.Q. Lim, S.A. Irvine, A. Agrawal, P.S. Mhaisalkar, J. An, C.K. Chua, S. Venkatraman, A novel 3D printing method for cell alignment and differentiation, Int. J. Bioprinting 1 (1) (2015) 57–65.
[26] P. Mozetic, S.M. Giannitelli, M. Gori, M. Trombetta, A. Rainer, Engineering muscle cell alignment through 3D bioprinting, J. Biomed. Mater. Res. A 105 (9) (2017) 2582–2588.
[27] E.D.P. Ker, A.S. Nain, L.E. Weiss, J. Wang, J. Suhan, C.H. Amon, P.G. Campbell, Bioprinting of growth factors onto aligned sub-micron fibrous scaffolds for simultaneous control of cell differentiation and alignment, Biomaterials 32 (32) (2011) 8097–8107.
[28] M. Costantini, S. Testa, P. Mozetic, A. Barbetta, C. Fuoco, E. Fornetti, F. Tamiro, S. Bernardini, J. Jaroszewicz, W. Swieszkowski, M. Trombetta, L. Castagnoli, D. Seliktar, P. Garstecki, G. Cesareni, S. Cannata, A. Rainer, C. Gargioli, Microfluidic-enhanced 3D bioprinting of aligned myoblast-laden hydrogels leads to functionally organized myofibers in vitro and in vivo, Biomaterials 131 (2017) 98–110.
[29] R. Langer, J. Vacanti, Tissue engineering, Science 260 (5110) (1993) 920–926.
[30] X.F. Cui, K. Breitenkamp, M.G. Finn, M. Lotz, D.D. D'Lima, Direct human cartilage repair using three-dimensional bioprinting technology, Tissue Eng. Part A 18 (11-12) (2012) 1304–1312.
[31] S. Michael, H. Sorg, C.T. Peck, L. Koch, A. Deiwick, B. Chichkov, P.M. Vogt, K. Reimers, Tissue engineered skin substitutes created by laser-assisted bioprinting form skin-like structures in the dorsal skin fold chamber in mice, PLoS One 8 (3) (2013).
[32] K. Tappa, U. Jammalamadaka, D.H. Ballard, T. Bruno, M.R. Israel, H. Vemula, J.M. Meacham, D.K. Mills, P.K. Woodard, J.A. Weisman, Medication eluting devices for the field of OBGYN (MEDOBGYN): 3D printed biodegradable hormone eluting constructs, a proof of concept study, PLoS One 12 (8) (2017) e0182929.
[33] P.R. Martinez, A. Goyanes, A.W. Basit, S. Gaisford, Fabrication of drug-loaded hydrogels with stereolithographic 3D printing, Int. J. Pharm. 532 (1) (2017) 313–317.
[34] N. Genina, J.P. Boetker, S. Colombo, N. Harmankaya, J. Rantanen, A. Bohr, Anti-tuberculosis drug combination for controlled oral delivery using 3D printed compartmental dosage forms: from drug product design to in vivo testing, J. Control. Release 268 (2017) 40–48.
[35] M.C. Du, B. Chen, Q.Y. Meng, S.M. Liu, X.F. Zheng, C. Zhang, H.R. Wang, H.Y. Li, N. Wang, J.W. Dai, 3D bioprinting of BMSC-laden methacrylamide gelatin scaffolds with CBD-BMP2-collagen microfibers, Biofabrication 7 (4) (2015).
[36] M.T. Poldervaart, H. Gremmels, K. van Deventer, J.O. Fledderus, F.C. Oner, M.C. Verhaar, W.J.A. Dhert, J. Alblas, Prolonged presence of VEGF promotes vascularization in 3D bioprinted scaffolds with defined architecture, J. Control. Release 184 (2014) 58–66.

Further reading

[37] A. Faulkner-Jones, C. Fyfe, D.J. Cornelissen, J. Gardner, J. King, A. Courtney, W.M. Shu, Bioprinting of human pluripotent stem cells and their directed differentiation into hepatocyte-like cells for the generation of mini-livers in 3D, Biofabrication 7 (4) (2015) 044102.

CHAPTER 6

Nanotechnology in improving medical devices for smart drug delivery

Ludwig Erik Aguilar
Bionanosystem Engineering, Chonbuk National University, Jeonju, Republic of Korea

Contents

6.1 Introduction

Biomedical devices have been used in the medical field since the dawn of modern medicine. These devices are used to augment, alleviate, and ease the process of healing for people who suffers from different medical conditions. Generally, any device used in the medical field can be classified as a biomedical device. It could range from invasive to noninvasive devices that monitor vital signs, prosthesis, and implants.

Nanotechnology has been in the forefront of science for at least three decades, it has created numerous breakthroughs and discovery in various fields across the sciences, as the great Professor Richard Feynman has said, "There's plenty of room at the bottom." Using nanotechnology to enhance

Biomimetic Nanoengineered Materials for Advanced Drug Delivery
https://doi.org/10.1016/B978-0-12-814944-7.00006-0

our lives is not just the one way to go but ultimately the best way to go. We are currently manipulating materials at its basic foundations, at scales that the natural world behaves differently. We can enhance and change the behaviors of the materials to improve their properties and nanotechnology is indeed a great tool if we want to have a bright future.

The biomedical field has been a beneficiary of nanotechnology. From nanoparticles for targeted delivery to nanotopology in tissue engineering, nanotechnology has been the tool of choice for researchers in improving the medical field. Although really promising, it is still at its infancy. The ultimate goal of nanomedicine is to solve the most dreaded problem in the world. Currently, it has been used to solve many problems, from mundane to complex, such as cancer therapy, tissue engineering, and regenerative therapy, and promising results has been documented.

In the field of medicine, biomedical implants by definition are man-made devices that augment the function of the body. The surface of implants that is in contact with the body might be made of a biomedical material such as titanium, silicone, or apatite depending on their intended function. It is the work of biomedical engineers and material scientists to create these devices to help alleviate or improve the quality of life of the patient, and as of now, nanotechnology has been on the forefront of research to improve the properties of biomedical implants and devices.

This chapter explores the benefits of using nanotechnology for modifying biomedical implants/devices to improve their properties in order to be able to deliver therapeutics in a controlled and on-demand manner. Nanofabrication techniques were employed to create new materials for drug delivery and also to enhance the ability of the said biomedical devices to be able to be used for curative purposes.

Since the biomedical device field encompasses broad varieties of materials and topics, we will just focus on two specific biomedical devices: first stents, further divided into two subtopics (vascular and non-vascular) and metallic bone implants. We will explore modifications made to these devices and how nanotechnology and material science can be employed for improving their performance. We will also explore how these modifications can be used in delivering therapeutics and how cells respond to different materials and conditions we create in line with the different drug delivery platforms we have added to these devices.

Surface modifications with the aim of delivering therapeutics is the main goal for this domain of research; we need to find a way to deliver therapeutics in a controlled manner to prolong the effect of the drug payload and also make sure that there are no toxicities involved. The prevailing idea is that with the use of polymers, we can achieve a controlled release of therapeutics, since it can be engineered in a manner that we can control its behavior so as to know which stimuli it can react upon. We can use external stimuli or the intrinsic dynamic conditions inside the body as the drug release triggering mechanism to control the release of different therapeutics from the medical devices/implant we are trying to modify. Stimuli-responsive polymers are great candidate materials for making sure that therapeutics can be loaded at a high loading capacity with an on-demand release capability. Only when specific conditions are met thus the drug payload can be delivered. These conditions can be adjusted depending on the chemical structure and doping materials used for creating these drug delivery platforms. We can classify the drug delivery method for biomedical devices as local drug delivery, in contrary to other drug delivery methods where multiple systems are involved, usually the device is at close proximity to the target tissue or organ, thus reducing the need for complicated chemical engineering for targeting ligands or thinking about metabolic impacts on the drug payload. So the main concern now is to have the optimal drug-loading capacity without compromising the intended function of the medical device involved.

6.2 Stents

Stents are metallic or polymeric mesh that has the ability to give off equal radial force to any bodily orifice to create patency and maintain normal function. There are various types of stents and are generally classified into two categories: (1) vascular and (2) non-vascular. Vascular stents as the name suggests are implanted or deployed in the vascular system. It could be used for angioplasty in the coronary system for maintaining normal blood perfusion to the heart muscle and also can be used in larger vessels to create normal hemodynamic flow that is usually disrupted by aneurysm located in the aorta, femoral artery, and neurovascular system. Non-vascular stents on the other hand are deployed in larger bodily orifice found in multiple systems in the body, such as the gastrointestinal system, respiratory system, and genitourinary system. This is due to the fact that non-vascular stents are meant to be deployed over cancerous or benign growths that occlude the orifices found in those systems. Since the nature of the occlusions is different for the said stents, the solutions for maintaining the patency of the stents to maintain its function are also different, although the problems encountered are still similar, such as stent restenosis/reocclusion or stent migration. In designing a proper modification for these stents certain criteria must be met. First, the drug-loading capability, it is essential to administer proper dosage to the target tissue to ensure therapeutic effects are met. Second, the cellular response to the polymeric drug platform being used. Third, the physical and chemical properties of the material of choice. These factors must be borne in mind when creating a drug delivery platform for either vascular or non-vascular stents.

6.2.1 Vascular stents

Let us put our focus first on vascular stents, when a patient develop irregular blood flow that leads to the disruption of normal oxygen perfusion to tissues, certain steps are evaluated by physicians to determine the proper intervention for the vasculature in question. There are three generations of vascular stents that had been developed in the recent decades. First-generation stents are made up of metallic mesh that physically corrects the occlusion by pushing the plaque or lipid buildup into the arterial wall to provide patency. For coronary arteries that are deemed to undergo angioplasty or percutaneous coronary intervention, first-generation stents are being placed to maintain patency and provide continuous blood flow to the heart muscles. But certain plaque buildup caused by lipid and platelet aggregation can occur after this intervention and could cause another obstruction in the luminal part of the stent rendering it useless. The patient has to undergo another surgery to correct this problem, which could alter their already fragile state of health. To avoid this problem, biomedical engineers have designed vascular stents having surface properties that could avoid plaque buildup in the luminal and abluminal parts of the stent by loading anti-thrombogenic drugs on the surface of the stents, these are called second-generation stents or drug eluting stents (DESs). Third-generation stents are bioresorbable and do not need any future intervention when the service life of the stent is finished. It would just be absorbed by the body over time by degrading into harmless ions or bioinert polymeric short chains. Candidates for third-generation stents are Mg alloys and bioresorbable polymers, for example, polycaprolactone (PCL), (PVDF, PVA, etc.). The development of vascular stents over time is described in Fig. 6.1.

How nanotechnology helps in this area of biomedical device depends on the use of nanocoating technology. Several papers have been published on the utilization of the properties of nanocoating to load specific molecules

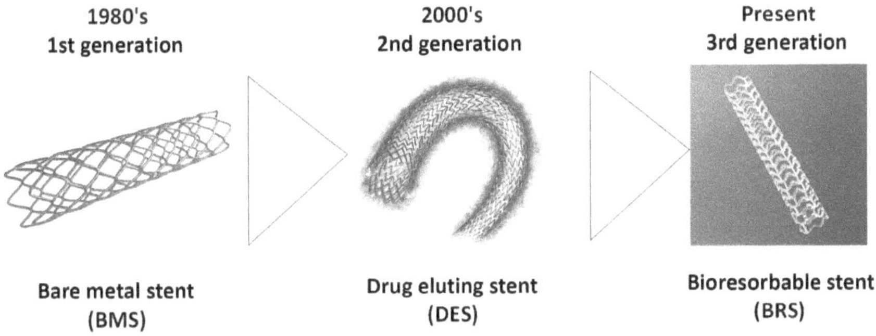

Fig. 6.1 Vascular stent design progression throughout the decades leading to the present time.

onto the surface of the metallic substrate. Researchers have used biopolymers such as polydopamine as an anchor to conjugate thiolated vascular endothelial growth factors (VEGFs) to stimulate re-endothelialization on the damage blood vessel caused by plaque buildup [1, 2]. The thickness of the coating plays a vital role in vascular stent technology as uncontrolled thickness often leads to inappropriate dosage, and stent mechanical property failures. Polymeric coatings that are too thick can also cause harmful effects as it can often break once inflated by the balloon catheter and can be a site for platelet buildup [3]. We have to make sure that the coating thickness is optimal and could not lead to cytotoxic response to cells. Other reports have demonstrated that thickness of about 20–100 nm is optimal for creating a functional layer in metallic substrate without sacrificing mechanical properties of the stent (Fig. 6.2).

6.2.2 Non-vascular stents

Non-vascular stents are similar to vascular stent in structure and function but applied elsewhere in the body. Metallic stents are made primarily for the vasculature system to maintain proper blood flow but eventually, engineers have developed designs such that the device can be applied to other parts of the body. Nowadays, stents are applied to malignant and benign strictures from the gastrointestinal, respiratory, and genitourinary systems. The development of non-vascular stents had also undergone changes over the past decades. Similar to vascular stents, reocclusion and migration are the main challenges it faces. Structural designs for these stents depend on the part of the body it is being applied to. Esophageal stents, for example, are bigger in diameter and longer in length, could also be easily removed when

Fig. 6.2 Surface coating of polydopamine onto the surface of a stent sample: (A and C) uncoated and (B and D) coated.

complications occur. The mechanical behavior must also be stronger than vascular stent, where self-expanding, shape memory materials are reliably used (e.g., Nitinol alloy). Nitinol is a biomedical grade alloy that maintains its shape and elasticity. It can also be programmed to maintain specific shapes when its geometry is manipulated and its Curie temperature reached. Nowadays, fully covered stents are used primarily to avoid stricture reocclusion inside the stent. These covers are made of polymeric materials (PU, PTFE) that have superb elasticity to address the mechanical requirements of non-vascular stenting. Much focus is laid on polymeric cover that has multipurpose properties than on non-vascular stents made of alloy material. First, provide a physical barrier for the tumor and the stent lumen and second, release therapeutics to reduce or maintain the tumor volume while the stent is in service.

Nanotechnology provides an answer to this problem by packaging therapeutics in polymeric membrane that has nanostructures, for example, polymeric covers that are made by nanofibers. This is done by electrospinning a polymer solution containing the drug payload onto the stent. Several works about this modification have been reported by these groups (cite). Electrospinning will be discussed further in the later part of the chapter. Polymers are great candidates since they provide different properties; they can be bioinert or bioactive depending on the design and purpose of the stent. Polymeric materials can also be nanoengineered to have higher tensile strength while having stimuli-responsive properties [4, 5].

Drug release from these polymers follows different mechanisms and behavior. Commercial non-vascular DES uses passive diffusion to release its drug payload. It is paramount to control the release of the drug to avoid over or under dosing the target tissue. Antiproliferative drugs such as paclitaxel (PTX), 5-Flourouracil (5-FU), and Gemcitabine had been studied to be used in such applications [6–8]. One advantage of using polymers for drug loading is that they have high drug-loading efficiency. Most high molecular weight polymers also have a high drug-loading capacity due to the increased polymeric chain length that can entrap the drug molecules.

6.3 Bone implants

Hard tissue (bone) implants are used to correct malformations and damages in the musculoskeletal system. There are several types of bone implants such as screws, plates, nails, sleeves, and wires depending on the site where augmentation is needed. The common denominator is the material used, and most common is titanium alloy and stainless steel. It has high mechanical strength, corrosion resistance, and superb biocompatibility that is why they are the material of choice when designing bone implants. The most common failure to be addressed with these biomedical devices is implant bacterial infection [9]. Having a device that could cause infection to the body can be detrimental to the patient and would need a second surgery to correct the damage caused by an infected implantation site. If the infection is too severe it could also lead to sepsis that can lead to a patient's death. To avoid this problem, biomedical engineers are creating a functional interfacial layer that can inhibit the growth of bacteria [10]. Coating techniques that serve as a boundary layer between the metallic and the biological interface are being developed to reduce the instances of implant infection. Antibacterial agents are being incorporated into these coatings to actively kill different types of Gram positive and Gram negative bacterial strains [11–16] such as silver ion- or silver nanoparticle-loaded coatings and flourine-doped nanopatterned coatings. These coatings are also engineered to have multifunctional properties (Fig. 6.3). They can also enhance osteogenic growth by manipulating

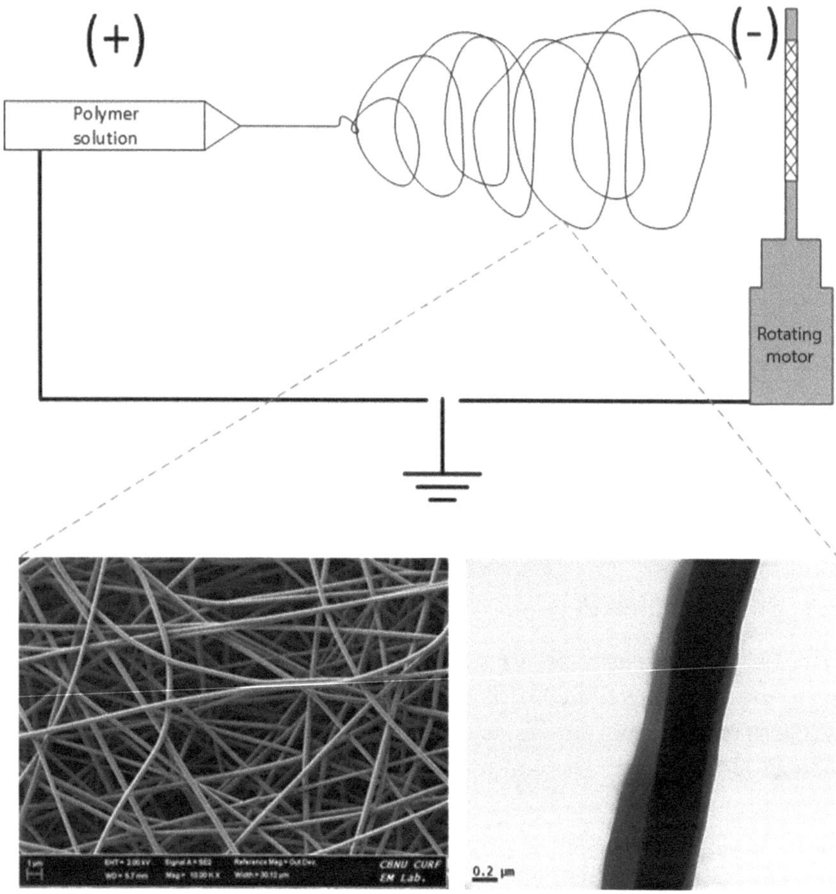

Fig. 6.3 Schematic illustration of how nanofibers can be incorporated as a stent cover and how nanofibers can be engineered to have multifunctional properties.

the nanotopography of the metallic substrate. Surface roughness of the implant can influence osteocytes to proliferate and differentiate on a faster rate compared to unmodified surfaces [13, 16].

6.4 Nanomodification techniques used for drug delivery in medical devices

Several techniques are available for modifying biomedical devices at the nanolevel, some of which were mentioned at the beginning of this chapter. We will now discuss in depth how they are applied to enhance the properties and effects of the previously discussed devices (stents and bone implants).

6.4.1 Electrospinning

Electrospinning is a technique for creating nanoscale polymeric fibers that utilize electric force to draw out the polymer solution that is in the liquid phase to fibers that are in solid phase. It is being extensively researched upon as a promising fabrication technique in various fields. In the biomedical field, it has several potential applications such as wound dressing material, cell tissue engineering, and drug delivery. Since it has a large surface area, it mimics the body's natural extracellular matrix and could help in tissue regeneration and could also release therapeutics by various mechanisms (diffusion, degradation, drug molecule-polymeric chain dissociation) [17].

In drug delivery, we can alter the property of the nanofiber by modifying the standard electrospinning setup. Coaxial electrospinning can combine two different polymeric materials so that the finished membrane has two distinct properties as shown in Fig. 6.4, a structure made of a shell and a core, this membrane will not sacrifice thickness for functionality because the electrospun polymers are combined into one fiber [18, 19]. Layer-by-layer approach can also be done during fabrication to have a multilayered biphasic functional material that could contain two or more therapeutics in one membrane but the resulting membrane thickness will be higher compared to the coaxial spun nanofiber. Two-nozzle approach can randomly combine two nanofibers that has different properties to achieve the same goal. There are several criteria for electrospinning polymeric materials. First, the molecular weight of the polymer must be within 1–10 kDa to achieve uniform

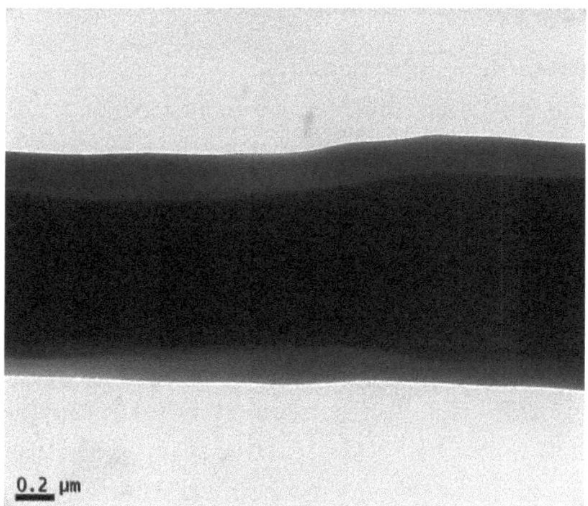

Fig. 6.4 Core-shell nanofiber structure with polyurethane core and poly(NIPAAm-co-HMAAm) shell.

diameter size distribution and to avoid beading. Fiber beading is one of the most common problem encountered when low molecular weight polymers are electrospun [17]. Due to the short polymeric chain in the solution, the electric drawing force overwhelms the short polymeric chain causing discontinued fiber formation leading to particle-/bead-like morphology. Most natural polymers like gelatin, chitosan, silk, and chitin are difficult to electrospun due to their insufficient molecular weight and shorter chain length (oligomers). In order to electrospin these materials, composites and blends with other polymers are created to ensure the formation of nanofibers. For stents, a combination of high mechanical strength and controlled drug release mechanism is desired to provide functionality and physical integrity, that is, delivery of therapeutics in an on-demand manner while maintaining its structural composition. A material that breaks upon deployment could mean another intervention and could only jeopardize the treatment. That is why composites that are combination of stimuli responsive polymers and reinforcing materials are being developed [4, 5]. PU is a reliable polymeric material for stent covers, due to its high elasticity and ease of manufacturing [20, 21]. Combining it with other polymers is a good strategy to create functional membranes that could withstand the rigors of and requirements of biomedical implants such as stents. PTFE is also a material that is currently used commercially but due to its ultrahigh molecular weight and chemical inertness, it is difficult to process using electrospinning [22–24].

6.4.2 Electrospray

Similar to electrospinning, electrospray uses electric force to draw out low molecular weight polymers that can be deposited onto a substrate. This creates a boundary layer that has distinct particle-like morphology. These particles can be loaded with drugs to be released in a controlled manner but due to low attachment to the substrate it is deemed impractical for other applications [7].

6.4.3 Electropolymerization coating technique

Electropolymerization uses electrochemistry to form polymers in the electrochemical cell by initiating a chain reaction from an electro radical. Monomers that are dispersed in the electrochemical solution can be reduced to polymeric forms. Also, polymers can be easily obtained by electrochemical oxidation of aromatic monomers on an electrode surface as a film state. To analyze the coating of polymers, we employ cyclic voltammetry (CV). It is a type of potentiodynamic electrochemical measurement. In a CV experiment, the working electrode potential is ramped linearly vs time. Unlike in linear sweep voltammetry, after the set potential is reached in a CV experiment, the working electrode's potential is ramped in the opposite direction to return to the initial potential. These cycles of ramps in potential may be repeated as many times as needed. The current at the working electrode is plotted vs the applied voltage (i.e., the working electrode's potential) to give the cyclic voltammogram trace. CV is generally used to study the electrochemical properties of an analyte in solution. A standard CV experiment employs a cell fitted with three electrodes: reference electrode, working electrode, and counter electrode. This combination is sometimes referred to as a three-electrode setup. Electrolyte is usually added to the sample solution to ensure sufficient conductivity. The solvent, electrolyte, and material composition of the working electrode determine the potential range that can be accessed during the experiment. The electrodes are immobile and sit in unstirred solutions during CV. This "still" solution method gives rise to CV's characteristic diffusion-controlled peaks. This method also allows a portion of the analyte to remain after reduction or oxidation so that it may display further redox activity. The stirring of the solution between CV traces is important in order to supply the electrode surface with fresh analyte for each new experiment. The solubility of an analyte can change drastically with its overall charge; as such it is common for reduced or oxidized analyte species to precipitate out onto the electrode. This layering of analyte can insulate the electrode surface, display its own redox activity in subsequent scans, or otherwise alter the electrode surface in a way that affects the CV measurements [25]. For this reason it is often necessary to clean the electrodes between scans. Common materials for the working electrode include glassy carbon, platinum, and gold. These electrodes are generally encased in a rod of inert insulator with a disk exposed at one end. A regular working electrode has a radius within an order of magnitude of 1 mm. A controlled surface area with a well-defined shape is necessary for being able to interpret CV results. To run CV experiments at very high scan rates a regular working electrode is insufficient. High scan rates create peaks with large currents and increased resistances, which result in distortions. Ultramicroelectrodes can be used to minimize the current and the resistance. The counter electrode, also known as the auxiliary or second electrode, can be any material that conducts current easily and will not react with the bulk solution. Reactions occurring at

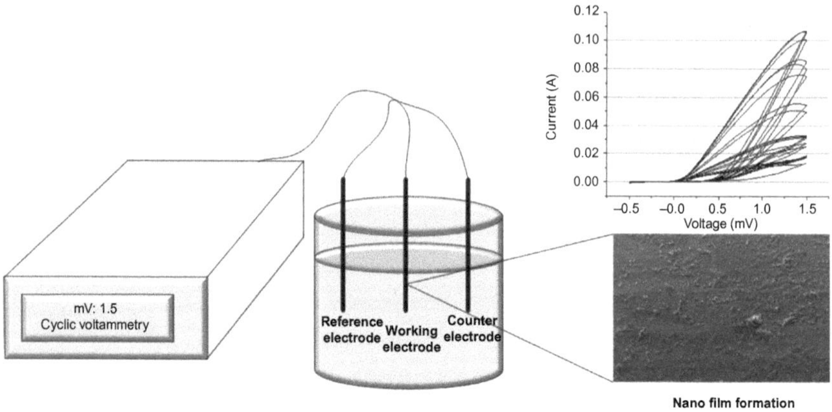

Fig. 6.5 Electropolymerization schematics of dopamine on a nitinol substrate.

the counter electrode surface are unimportant as long as it continues to conduct current well. To maintain the observed current the counter electrode will often oxidize or reduce the solvent or bulk electrolyte.

Some phenolics or phenolic-like molecules (e.g., dopamine, gallic acid, vanillin, etc.) have the natural tendency to create nanofilms on the surface of the electrodes [26, 27]. This a natural occurrence that can be utilized as a coating technique for conductive materials. If we place, for example, a stent as the working electrode, the monomer in the electrochemical cell will polymerize onto the surface of the stent and can be used as an interfacial layer as shown in Fig. 6.5. Some phenols have intrinsic anticancer or anti-inflammatory properties, and if the phenol molecule has a catechol moiety, we can also conjugate thiolated drugs or selected types of chemotherapeutics (bortezomib, cisplatin) via Michael addition chemistry.

6.5 Stimuli-responsive polymers for biomedical device drug delivery

Suitable materials must be considered when creating a drug delivery platform for biomedical devices. It must be able to deliver its drug payload without compromising the intended function of the device. The drug payload must also compliment the device by prolonging and enhancing its properties. A great material for drug delivery must be able to release its payload in a well-controlled manner; this can be achieved by using polymeric materials that can react with its environment via changes in pH, temperature, or biochemical cues. We call these kinds of polymers stimuli-responsive polymers;

they have the capability to undergo changes in phases, wettability, or stability depending on the condition of the environment. The human body is a dynamic place, due to homeostasis; it can regulate its temperature, pH, and molecular balance and most often than not they are not fixed. We can exploit these changes by creating stimuli-responsive polymeric drug platforms and incorporating them into our biomedical devices.

6.5.1 Temperature-responsive polymers

One example of temperature-responsive polymer is poly(*N*-isopropylacrylamide) or p-(NIPAAm). It has the ability to change its physical state from hydrophilic to hydrophobic due to its lower critical solubility temperature (LCST) property, p-(NIPAAm) has an LCST point of 32°C, much lower than the normal body temperature (Fig. 6.6). Because of this, we need to copolymerize it with a hydrophilic monomer to increase its LCST property [28, 29]. Depending on the ratio of the copolymer we can tailor its LCST temperature to a much higher range, depending on the intended application [4, 30]. The lower the ratio of p-(NIPAAm) in the copolymer the higher the LCST transition phase, but if the ratio has reached its limit, the LCST property of the copolymer can be lost. Therefore, it is important when designing a copolymer that we know the optimal ratio of the copolymers. When we incorporate a drug into the copolymer network, the drug payload is maintained when the temperature of the system is below its LCST point. When we increase the temperature of the copolymer to its LCST point, the

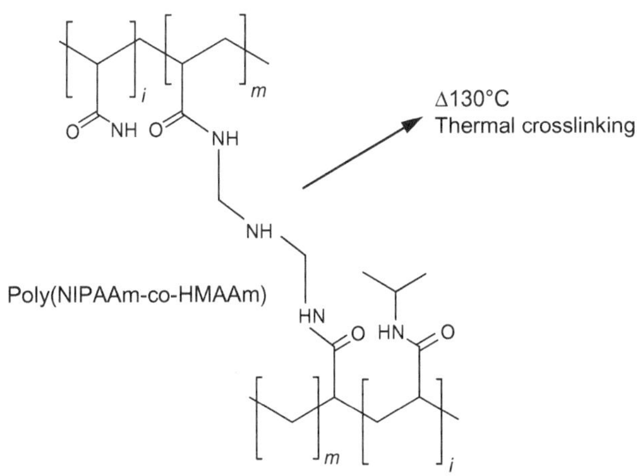

Fig. 6.6 Chemical structure of poly(NIPAAm) copolymerized with *N*′-hydroxylmethyl methacrylate to increase its LCST property.

copolymer becomes hydrophobic and releases its drug payload. It is similar to a sponge squeezing out its water content when pressure is applied, the copolymer network collapses when the temperature of the system reaches its LCST point [31].

6.5.2 pH-responsive polymers

pH-responsive polymers can react to the changes in pH of its surrounding by changing its dimensions. It can be explained by the presence of certain functional groups in the polymeric network that are sensitive to pH. These polymers can swell, degrade, or collapse depending on the pH of the media. pH-sensitive polymers can be responsive to either basic or acidic changes in the system [5, 29, 32]. In drug delivery platforms, polydopamine is a widely used polymer [1, 33, 34]. Dopamine can be seen inside the body as a naturally occurring phenol. It is a neurotransmitter responsible for reward-motivated behavior. Dopamine also has the ability to self-polymerize under alkaline conditions making it very versatile as a coating agent. The polymerized form of dopamine has the ability to adhere onto the surface of biomedical devices due to its catechol functional groups. The catechol groups on polydopamine can also be a conjugation site for metal-containing drugs or thiolated growth factors [33, 35–38]. The link between the drug and the coating is pH sensitive and only in the presence of acidic pH the conjugation could be dissociated and the drug payload be released. This is important because we can use the intrinsic acidic pH found in the body, that is, in the stomach and duodenum, as drug delivery triggering mechanism [5, 34]. Also, it was proved that we can use cancer physiology as an acidic environment-driven drug delivery triggering mechanism due to increased glycolytic activity and hypoxic metabolism of its cells (Fig. 6.7).

6.6 Conclusion

Several techniques can be utilized to serve as drug delivery platforms for biomedical devices. As discussed in this chapter, electrospinning can be used as an effective nano package for therapeutics. Nanofibers can be tailored to specific morphology to incorporate different properties needed for biomedical device application. Also, nanocoatings using electropolymerization can provide us a high degree of control when coating conductive biomedical devices such as stents, and titanium implants. The coating can serve as another biocompatible surface and drug delivery platform to reduce the complications that are frequently encountered with biomedical devices.

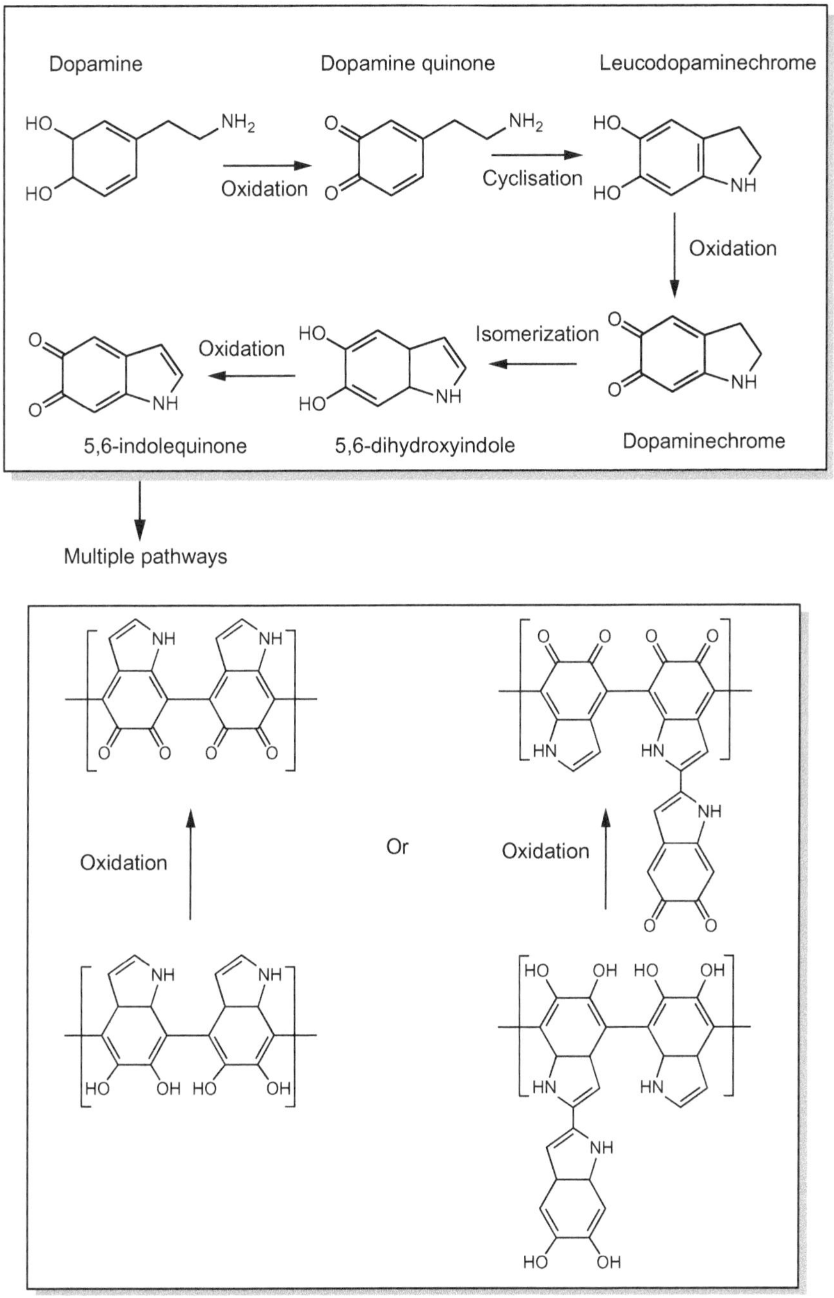

Fig. 6.7 Polymerization of dopamine. Schematic illustration of multiple pathways and polymeric structures of polydopamine.

With these techniques, stimuli-responsive polymers are the best candidates as the material of choice due to their capability to release their payload in a controlled manner. Further research is needed to add to the increasing demand of the field. Better fabrication techniques are needed so that we can scale-up the production of biomedical devices that has multiple properties.

References

[1] J.L. Wang, et al., Electropolymerization of dopamine for surface modification of complex-shaped cardiovascular stents, Biomaterials 35 (27) (2014) 7679–7689.
[2] Y. Wei, et al., Surface engineering of cardiovascular stent with endothelial cell selectivity for in vivo re-endothelialisation, Biomaterials 34 (11) (2013) 2588–2599.
[3] Y. Farhatnia, et al., Next generation covered stents made from nanocomposite materials: a complete assessment of uniformity, integrity and biomechanical properties, Nanomedicine 12 (1) (2016) 1–12.
[4] L.E. Aguilar, et al., On-demand drug release and hyperthermia therapy applications of thermoresponsive poly-(NIPAAm-co-HMAAm)/polyurethane core-shell nanofiber mat on non-vascular nitinol stents, Nanomedicine 13 (2) (2017) 527–538.
[5] L.E. Aguilar, et al., Electrospun polyurethane/Eudragit (R) L100-55 composite mats for the pH dependent release of paclitaxel on duodenal stent cover application, Int. J. Pharm. 478 (1) (2015) 1–8.
[6] J. Liu, et al., Paclitaxel or 5-fluorouracil/esophageal stent combinations as a novel approach for the treatment of esophageal cancer, Biomaterials 53 (Suppl. C) (2015) 592–599.
[7] L. Zhao, et al., Rational design of drug-eluting stents via electrospray and in vivo evaluation of preventing oesophageal stricture, RSC Adv. 4 (32) (2014) 16885–16892.
[8] J.W. Lee, S.-G. Yang, K. Na, Gemcitabine-releasing polymeric films for covered self-expandable metallic stent in treatment of gastrointestinal cancer, Int. J. Pharm. 427 (2) (2012) 276–283.
[9] J. Mouhyi, D.M. Dohan Ehrenfest, T. Albrektsson, The peri-implantitis: implant surfaces, microstructure, and physicochemical aspects, Clin. Implant. Dent. Relat. Res. 14 (2) (2012) 170–183.
[10] J. Qu, et al., Silver/hydroxyapatite composite coatings on porous titanium surfaces by sol-gel method, J. Biomed. Mater. Res. B Appl. Biomater. 97B (1) (2011) 40–48.
[11] C. Baker, et al., Synthesis and antibacterial properties of silver nanoparticles, J. Nanosci. Nanotechnol. 5 (2) (2005) 244–249.
[12] V. Ball, et al., The reduction of Ag+ in metallic silver on pseudomelanin films allows for antibacterial activity but does not imply unpaired electrons, J. Colloid Interface Sci. 364 (2) (2011) 359–365.
[13] A. GhavamiNejad, et al., Immobilization of silver nanoparticles on electropolymerized polydopamine films for metal implant applications, Colloid Interface Sci. Commun. 6 (2015) 5–8.
[14] A. Panáček, et al., Silver colloid nanoparticles: synthesis, characterization, and their antibacterial activity, J. Phys. Chem. B 110 (33) (2006) 16248–16253.
[15] M. Sureshkumar, P.-N. Lee, C.-K. Lee, Stepwise assembly of multimetallic nanoparticlesvia self-polymerized polydopamine, J. Mater. Chem. 21 (33) (2011) 12316–12320.

[16] J.H. Zhou, et al., F-Doped micropore/nanorod hierarchically patterned coatings for improving antibacterial and osteogenic activities of bone implants in bacteria-infected cases, ACS Biomater. Sci. Eng. 3 (7) (2017) 1437–1450.
[17] S. Agarwal, A. Greiner, J.H. Wendorff, Electrospinning of manmade and biopolymer nanofibers-progress in techniques, materials, and applications, Adv. Funct. Mater. 19 (18) (2009) 2863–2879.
[18] A.K. Moghe, B.S. Gupta, Co-axial electrospinning for nanofiber structures: preparation and applications, Polym. Rev. 48 (2) (2008) 353–377.
[19] H.L. Qu, S.Y. Wei, Z.H. Guo, Coaxial electrospun nanostructures and their applications, J. Mater. Chem. A 1 (38) (2013) 11513–11528.
[20] D. Jeong, et al., Nonvascular drug-eluting stent coated with sodium caprate-incorporated polyurethane for the efficient penetration of paclitaxel into tumor tissue, J. Biomater. Appl. 29 (8) (2015) 1133–1144.
[21] J.I. Kim, et al., Electrospun propolis/polyurethane composite nanofibers for biomedical applications, Mater. Sci. Eng. C Mater. Biol. Appl. 44 (2014) 52–57.
[22] S. Moon, S.G. Yang, K. Na, An acetylated polysaccharide-PTFE membrane-covered stent for the delivery of gemcitabine for treatment of gastrointestinal cancer and related stenosis, Biomaterials 32 (14) (2011) 3603–3610.
[23] T.H. Park, E.A. Jo, K. Na, Development of polymeric coating material for effective drug-eluting stent, Polymer 35 (5) (2011) 483–487.
[24] W.K. Wang, et al., Connexin 43 gene therapy delivered by polymer-modified salmonella in murine tumor models, Polymers 6 (3) (2014) 1119–1128.
[25] Z.H. Wang, et al., Voltammetric determination of salicylic acid by molecularly imprinted film modified electrodes, Int. J. Polym. Anal. Charact. 17 (2) (2012) 122–132.
[26] R. Okner, et al., Electropolymerized tricopolymer based on N-pyrrole derivatives as a primer coating for improving the performance of a drug-eluting stent, ACS Appl. Mater. Interfaces 1 (4) (2009) 758–767.
[27] B. Stöckle, et al., Precise control of polydopamine film formation by electropolymerization, Macromol. Symp. 346 (1) (2014) 73–81.
[28] B. Kim, D. Hong, W.V. Chang, LCST and UCST double-phase transitions of poly (N-isopropylacrylamide-co-2-acrylamidoglycolic acid)/poly(dimethylaminoethyl methacrylate) complex, Colloid Polym. Sci. 293 (3) (2015) 699–708.
[29] S.P. Rwei, et al., Thermo- and pH- responsive copolymers: poly(N-isopropylacrylamide-co-IAM) copolymers, J. Appl. Polym. Sci. 132 (33) (2015).
[30] J.W. Seo, U.S. Shin, Cationic effect of an ionic copolymer with a temperature-responsive charateristic on the LCST value: a broad LCST spectrum of 35 to 46 A degrees C, Macromol. Res. 23 (11) (2015) 1073–1078.
[31] P. Techawanitchai, et al., Surface design with self-heating smart polymers for on-off switchable traps, Sci. Technol. Adv. Mater. 12 (4) (2011).
[32] C.L. He, et al., Stimuli-sensitive synthetic polypeptide-based materials for drug and gene delivery, Adv. Healthc. Mater. 1 (1) (2012) 48–78.
[33] A.R.K. Sasikala, et al., A smart magnetic nanoplatform for synergistic anticancer therapy: manoeuvring mussel-inspired functional magnetic nanoparticles for pH responsive anticancer drug delivery and hyperthermia, Nanoscale 7 (43) (2015) 18119–18128.
[34] L.E. Aguilar, et al., Functionalized non-vascular nitinol stent via electropolymerized polydopamine thin film coating loaded with bortezomib adjunct to hyperthermia therapy, Sci. Rep. 7 (1) (2017) 9432.
[35] A. GhavamiNejad, et al., Mussel-inspired electrospun smart magnetic nanofibers for hyperthermic chemotherapy, Adv. Funct. Mater. 25 (19) (2015) 2867–2875.

[36] A.R.K. Sasikala, et al., An implantable smart magnetic nanofiber device for endoscopic hyperthermia treatment and tumor-triggered controlled drug release, Acta Biomater. 31 (2016) 122–133.
[37] J. Su, et al., Catechol polymers for pH-responsive, targeted drug delivery to cancer cells, J. Am. Chem. Soc. 133 (31) (2011) 11850–11853.
[38] S. Wu, et al., pH-Responsive drug delivery by amphiphilic copolymer through boronate–catechol complexation, ChemPlusChem 78 (2) (2013) 175–184.

CHAPTER 7

Radiofrequency-sensitive nanocarriers for cancer drug delivery

N. Sanoj Rejinold, Yeu Chun Kim
Department of Chemical and Biomolecular Engineering, Korea Advanced Institute of Science and Technology (KAIST), Daejeon, Republic of Korea

Contents

7.1 Introduction

"Cancer" remains an incurable disease, killing many lives all around the world. Being a deadly disease, it can affect any part of the body causing severe health issues resulting in fatal death [1–6]. Recent developments in cancer therapy led to many changes particularly in leading radiation research area, which utilizes radiofrequency (RF) ablation (RFA). The RFA being a nonhazardous treatment has attracted the medical field since the early 1990s [7–9]. The RFA make use of RF current (10 kHz and 900 MHz) to ablate cancer cells very efficiently [10]. Compared to other radiation techniques, such as radio and laser ablation therapies, the great infiltration capacity of RF waves enables better cancer cell-killing effect. The RF power is affected by the frequency of electromagnetic waves, tissue penetrability, etc. [11–14].

Biomimetic Nanoengineered Materials for Advanced Drug Delivery
https://doi.org/10.1016/B978-0-12-814944-7.00007-2

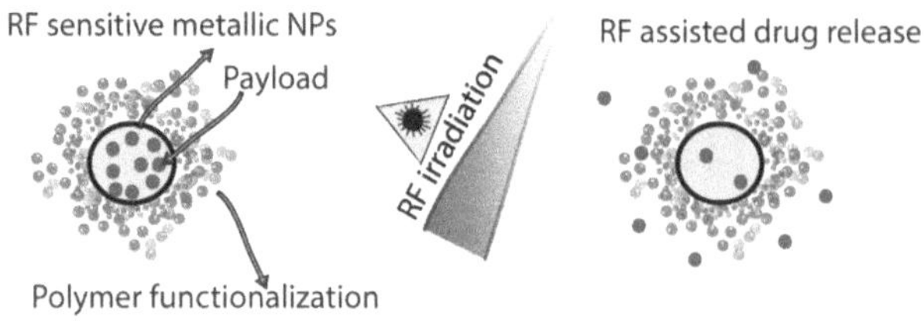

Fig. 7.1 RF-assisted drug release strategy.

Generally, deep-seated malignancies are difficult to treat using conventional therapy, whereas the FDA (Food and Drug Administration)-approved RFA would be highly beneficial because of their nonionizing and high penetrability. RFA generates heat through molecular ionic agitations when they are adjusted with the alternating current as it pass through the malignant tissues [14].

These interesting features of RFA excited many scientists to use it for biomedical applications especially to treat cancers. Thus, metal-based NPs such as gold (Au) and iron (Fe_3O_4) have been widely explored in combination with polymers for drug delivery applications. Usually, metallic nanoparticles (MNPs) alone or their composite nanocarriers are preferred so that the loaded/encapsulated drug molecules can be easily delivered in a controlled way with respect to the external RF application [15]. The general concept of RF-sensitive nanocarriers is shown in Fig. 7.1.

Usually, RF-sensitive Au or Fe_3O_4 NPs are explored as the base material, which is then functionalized with polymers that can hold the payload well. On exposure to RF radiation, the RF-sensitive nanomaterials will heat the polymer, releasing the payloads in a controlled way [16].

This chapter deals with recent approaches in RF-based drug delivery to treat cancer cells in detail with applications and future directions.

7.2 RF-sensitive therapeutic nanocarriers for cancer nanotherapy

A nanostructure can become RF responsive only when it can heat up in an external RF field. Ideally, Au, iron oxide, copper, carbon nanomaterials, quantum dots (QDs), and silicon NPs have been recognized as RF-sensitive therapeutic nanostructures in the in vitro experiments (Fig. 7.2). Both Joule heating and magnetic heating have been reported to be the basic mechanism involved in the RF-assisted heating. It has been reported that effective RF heating conditions are in the range of 40–600 W [17–21].

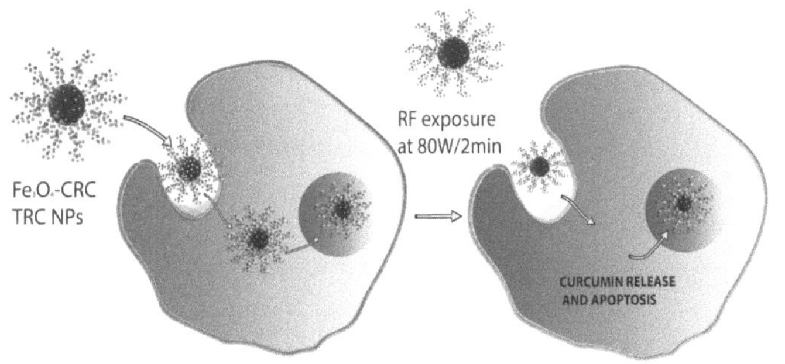

Fig. 7.2 RF-assisted CRC delivery to cancer cells from Fe_3O_4-CRC-TRC NPs under low RF exposure at 80 W/2 min.

Since an RF nanocarrier loads or encapsulates the payload in the RF-sensitive nanomaterial, it has to associate with polymer functionalized Au or Fe_3O_4 NPs. The coated or functionalized polymer should be temperature sensitive as well [14]. Commonly used polymers for such functionalization are poly(*N*-vinyl caprolactam) and poly(*N*-isopropyl acrylamide). In addition to the MNPs, the liposomes have also been found to be very much RF-induced heat sensitive [22].

Based on the core material, the RF-sensitive nanocarriers are classified into two groups: (1) metallic nanocarrier and (2) nonmetallic nanocarrier. MNPs such as Au, Fe_3O_4, Tin, etc., have been used as core RF-sensitive materials for cancer drug delivery applications. The nonmetallic carriers include mainly liposomes and carbon-based nanocarriers.

7.3 Gold-based RF-sensitive nanocarriers

There have been various studies on RF-aided oncology practices using Au-NPs [23–25]. Many researchers have reported RF-aided heatability of Au-NPs and their potential ability to kill cancer cells effectively [26–35]. The heating potential of Au-NPs RF was first found by Kanzius [36]. He observed that Au-NPs when excited with RF waves could effectively generate therapeutic heat, to "cook" malignant cells either via apoptotic or necrotic pathway [37].

The Kanzus device use antibody-conjugated Au-NPs, which are directly delivered to the patient intravenously. When the injected NPs reach the desired tumor site, the patient is subject to an RF source (which is applied externally) [36]. Therapeutic heat will be generated in this manner, which can then have "intratumoral hyperthermia," in either apoptotic or necrotic manner without causing any damage to the normal healthy cells. Compared to the existing radiotherapies, this method has no or negligible toxicity. The technique literally needs only an RF source and Au-NPs. The RF technology have already proven to be safe as it has major applications in cardio diseases, while Au-NPs have the potential to treat rheumatoid arthritis, with no toxicity [36, 38]. Kanzius RF device can generate radio waves to induce noninvasive cell toxicity to the targeted malignancies/tissues effectively at 13.56 MHz, safe for human use [39].

Before injecting into the body, tumor-targeting antibody is tagged with Au-NPs to enhance uptake and therapeutic effect [39]. The patient is then given these functionalized Au-NPs and because of its site specificity, and selectively, the Au-NPs can have no lethal effects on the healthy cells. Therapeutic heat will be generated only when an external RF source is supplied.

Another study has reported biopolymer-based composite RF sensitive nanogels for anticancer therapy. Biopolymer chitin was functionalized with Au-NPs/MnO_2 nanorods to make an RF-sensitive therapeutic platform "ACM-TNGs." Under optimum RF condition, these nanotherapeutics were able to eradicate mammary gland tumor cells effectively. The study showed that modified Au-NPs enhance RF sensitivity as composite NGs. The NP concentrations were minimized to get maximum RF response under 20–100 W. The therapeutic heat was found to be directly proportional to the concentration of MnO_2 nanorods. When T47D was treated with ACM-TNGs at RF exposure of 100 W for 2 min, they were eradicated effectively [40].

7.4 MNPs as RF-sensitive nanocarriers

MNPs have been recognized as smart RF therapeutic system for treating malignancies related to cancers [41] with FDA recognition as contrast agents [42–46]. They are biocompatible [47], bioresorbable [48, 49], and susceptible to modification. Even though, they are predominantly used as contrasting agents, their RF therapeutic potential has not yet been fully established. Unlike Au-NPs, the MNPs can have RF heating through magnetic moment [36].

Chandraprabha et al. studied hyperthermic properties of various MNPs coated with starch, glycine, and sodium oleate under external magnetic field, resulting in therapeutic heat of 38–55°C. The critical point was the concentration of MNPs for heat generation under a magnetic field. Compared to other coatings, both starch and glycine improved magnetic hyperthermia of the MNPs, indicating that sodium oleate coated MNPs are not a good option for magnetic hyperthermia [50].

Similarly, MNPs for RF application have been synthesized via a green route reported by Rejinold et al. *O*-carboxy methyl chitosan was used to make ~80 nm-sized MNPs. The optimum RF conditions were chosen (80 W for 2 min) in order to minimize background ion or molecule heating in vivo [51]. Other studies showed that once these MNs were combined with thermosensitive polymer called chitosan-*graft*-poly(*N*-vinyl caprolactam), they were able to deliver CRC in vivo effectively. At an 80 W 2-min RF exposure, the MNP-loaded NPs were able to attain the LCST (42°C) of the thermosensitive polymer, enabling a sustained CRC release and causing 4T1 breast cancer cells to undergo apoptosis. These results proved that MNP-modified NPs system could serve as improved magnetic hyperthermic agents for treating various cancers (Fig. 7.2) [16].

Thikra et al. have studied MNPs of various sizes to understand their magnetic hyperthermic heating capacity and pyro toxicity on RF excitation. Among other samples 25 nm-sized MNPs showed maximum effect at low-frequency RF irradiation (350 kHz). Small-sized (15 nm) MNPs had a maximum effect at higher frequency (13.56 MHz). These effects were concentration-dependent leading to enhanced therapeutic effects on MCF-7 cells via reactive oxygen species (ROS)-mediated apoptosis as visualized by SEM (scanning electron microscope) and TEM (transmission electron microscope) [52]. Another study explored the nanocubes of ferrimagnetic iron oxide as a magnetic hyperthermic platform stabilized by chitosan. The in vivo study showed that these NPs could effectively eradicate the targeted A549 malignant cells by caspase-mediated apoptosis [53] (Fig. 7.3).

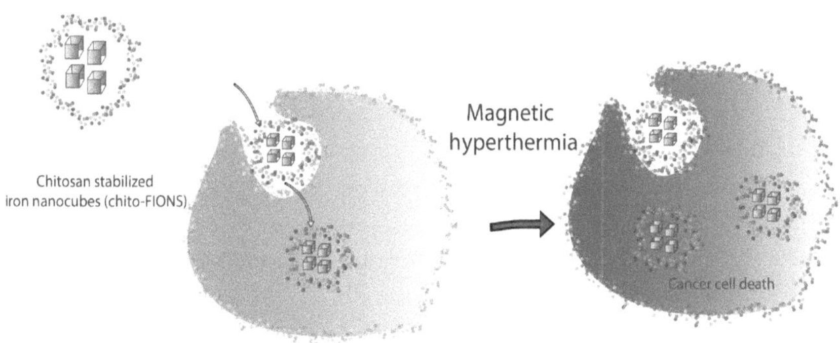

Fig. 7.3 Cancer Magnetic hyperthermia generated by the chitosan-DOPA-modified iron oxide nanocubes at in vitro conditions.

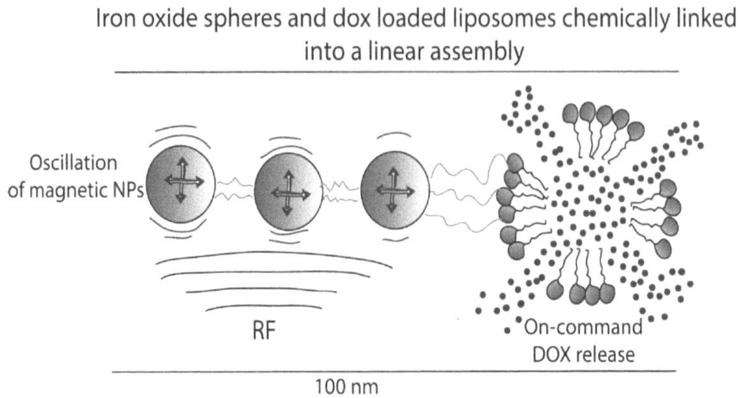

Fig. 7.4 On-command RF-assisted DOX release from iron oxide sphere-liposome linear assembly.

Peiris et al. have reported multicomponent nanochains with loaded drug to have intratumoral delivery with specific targetability. The NP contains triple chains of magnetic nanospheres and liposomal doxorubicin assembled with the nanochain (100 nm). (Fig. 7.4) These NPs had high blood circulation with enhanced intratumoral localization in vivo. The developed NPs act as a mechanical transducer transferring RF energy to liposomes that have drug in it. The destabilized liposome structures could easily have triggered drug transfer to intratumoral areas resulting in improved antitumor effects [54].

Karmakar et al. have demonstrated that by modifying NPs with antibodies, nonselective uptake can be reduced greatly, for example, epidermal growth factor (EGF)-modified NPs. EGF-conjugated carbon-coated iron oxide magnetic NPs (C/Fe MNPs) had improved accumulation in EGF expressing tumor cells after modification with EGF. A short-time RF excitation damaged the NP-treated MCF-7 cells with $<97\%$ toxicity while the unmodified NPs did not have any effects on cancer cells. The thermocytotoxicity is induced through a caspase-3-deficient pathway. These kinds of conjugation strategy can not only improve the selectivity, but also enhance the biocompatibility on MCF-7 cells compared to control samples [55].

Few-layered C/FeMNPs of size 7–9 nm are loaded with doxorubicin, and erlotinib, by stacking onto the carbon shells. The drug release was triggered by either RF excitation or pH variation. These multidrug-loaded NP systems showed synergistic anticancer effect by chemo and thermal toxicity to pancreatic cancer cells efficiently. This finding throws light on the fact that these C/Fe MNPs could be beneficial for multidrug loading and thermotherapy to treat various cancer types [56].

Kong et al. studied multifunctional magnetic nanoplatform where magnetic NPs coencapsulated with chemo agents have a magnetic responsive therapeutic outcome. These NPs successfully abolished tumors through a well-controlled on-off switchable drug release on RF exposure. The nanocapsules showed 80–150 nm even after co-loading and internalized very well intratumoral in mammary tumor model (mouse) without any further proliferation [57].

7.5 QD-based RF-sensitive nanocarriers

Glazer et al. came up with the idea that QDs could act as a promising RFsensitive nanotherapeutics to treat pancreatic cancer cells efficiently. When modified with specific antibody, targeted tumor will have maximum intratumoral heating which is very specific and selective.

EGFR targeting cetuximab (C225) monoclonal antibody was conjugated with Cd/Se- and In/GaP (phosphide) QDs in order to have a selective intracellular RF-assisted thermal cytotoxicity. The RF-assisted heating was concentration-dependent in vitro. A selective RF heating was noted on the cocultures of human malignant cells with varied EGFR-1 expressions. The targeted QDs showed selective toxicity to EGFR expressing pancreatic tumor cells than the lower EGFR expressing mammary tumor cells [58]. These studies have highlighted the theranostic properties of targeted QDs for simultaneous intratumoral imaging and RF-assisted pyro toxicity making them ideal for cancer drug delivery field.

Superparamagnetic biocompatible luminescent nanocomposites of Fe_3O_4 and SiO_2-QDs (IQ) have been reported by Yang et al. Silica was initially shelled on Fe_3O_4-NPs to form Fe_3O_4/SiO_2 (IOS), then modified with hydrophilic CdSe-ZnS QDs and thiol conjugation. Multilayered nanocomposites could improve RF-assisted therapeutic heating and intratumoral imaging. These nanostructures showed potential superparamagnetic and photoluminescent characteristics as well. IOS NPs were successfully localized into Panc-1 within 24 h incubation. An amount of 1.66 μg/mL of IOS NPs showed excellent RF-assisted cytotoxicity toward Panc-1 cells (99.2% cell death) within 2-min RF exposure. Precisely they had high magnetic properties and self-heating characteristics with excellent cytocompatibility. These outstanding results proved that the IOS NPs could be effective as RF-sensitive nanotherapeutics for cancer treatment [59].

7.6 Carbon-based RF-sensitive nanocarriers

Single-walled carbon nanotubes (SWNTs) are known for their outstanding physicochemical properties, which enables them to possess various biomedical applications [60–63]. RF heating capability of SWNTs was reported in 2007 [64]. Gannon et al. reported that cancer cell-targeted SWNTs could induce thermal cytotoxicity under mild RF conditions. The targeted approach improves the probability of injected NPs to reach specifically into the cancer cells. The authors have studied about CoMoCAT® SWNTs modified with Kentera, which is polyphenylene ethylene (PPE). PPE modification on SWNTs is achieved via non-covalent π–π stacking. The functionalized SWNTs showed larger RF therapeutic heating with low concentration of 5 mg/L with improved thermo-cytotoxicity on Hep3B, HepG2, and Panc-1 under 800 W/2 min RF excitation. The RF-induced cytotoxicity was concentration-dependent and 100% toxicity was found at 500 mg/L in all the three tested cell lines [64].

Graphene has been found to be useful for RF-aided cancer therapy as well. Under RF, the transferrin functionalized graphene NPs induced thermo cytotoxicity and ablated drug- or radiation-resistant cancer cells efficiently [65]. Similarly, transferrin conjugation on graphene NPs improved targeted RF hyperthermia on hepatic carcinoma cells as well [66].

7.7 Cobalt-based NPs for RF cancer therapy

Xu et al. have studied about the RF therapeutic benefits of cobalt NPs (C-Co-NPs) [37]. Cubic crystalline graphitic carbon (7 nm)-modified ferromagnetic Co-NPs (C-Co-NPs) efficiently generated nano hyperthermic RF heat. The Co-NPs were conserved within carbon shell in its metallic form as confirmed by XPES. Fluorescence microscopy and Raman spectroscopy confirmed their cellular and subcellular penetration and they induced therapeutic heat, cooking tumor cells effectively under 350 kHz of lower RF pulses. The whole process was RF power as well as concentration dependent. The C-Co-NPs showed very high specificity for RF absorption and heating than the SWNTs. C-Co-NPs treated cells had DNA as well as nuclear deformation and membrane disintegration on HeLa cells, showing complete eradication of tumor cells. These results prove that C-Co-NPs could serve as an excellent in vivo RF therapeutic agent, and additionally kill bacteria.

7.8 Tin-based NPs as RF-sensitive nanocarriers

Though RFA and chemotherapy were separately approved for liver cancer treatment, there are many limitations for both strategies; for example, non-uniform and nonspecific heating for RFA and systemic side effects make them having poor therapeutic side effects. These problems could be solved by a co-loading approach where RF heatable NPs and DOX will be encapsulated in biodegradable NPs. For this, Tin(II) ions and DOX were co-loaded in alginate NPs to have simultaneous chemo and hyperthermia under mild RF conditions. In order to make them more selective toward the liver cancer cells, asialo-glycoprotein targeting galactose was surface functionalized. Targeted NPs improved uptake and thermo-cytotoxicity in liver cancer cells. In vivo RFA was ~44% as demonstrated by magnetic resonance imaging along with RF-controlled Dox release (Fig. 7.4). Approximately 20°C RF thermal response was found at the tumor site in orthotopic rat liver tumor models resulting in uniform augmented ablation (~80%) at mild RF exposure of 15 W for 1 min. Since alginate are biocompatible and degradable, it could serve as an excellent translational agent for combined RF hyperthermia along with triggered DOX release [67] (Fig. 7.5).

7.9 Liposomes

Liposomes have been known for its thermal sensitivity [68, 69]. Liposomal doxorubicin and quercetin were used along with RFA to understand how modulating heat shock protein (HSP) affects tumor coagulation in animal

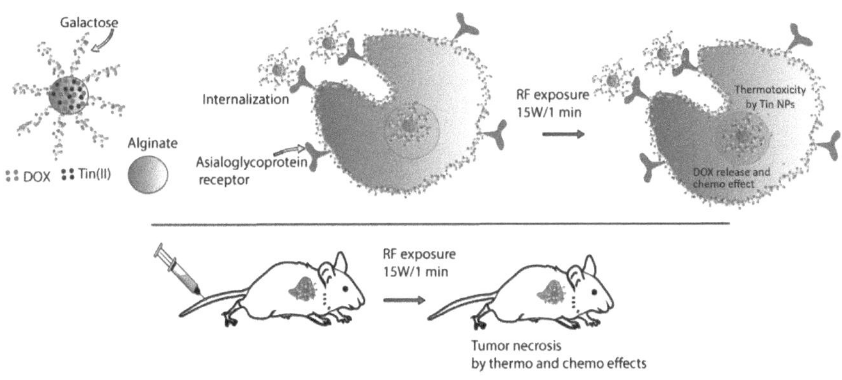

Fig. 7.5 Tin-doped alginate/galactose NPs for targeted hyperthermia and chemotherapy delivery to hepatic cancer cells [67].

cancer model. This study concluded that suppression of HSP amounts using liposomal quercetin formulation could enhance the RF-induced apoptosis greatly. The tumor coagulation is further enhanced by antitumor adjuvant liposomal doxorubicin [70].

Further study by combining RFA with an injectable liposomal HSP showed a reduced tumor growth with a significant survival rate in a small animal tumor model. This study focussed on understanding the effect of injectable liposomal quercetin formulation alone and with doxorubicin liposomal formulation on tumor growth and survival rate in combination with RFA in a rat tumor model [71]. The results showed that combination of IV liposomal quercetin and RFA had excellent antitumor efficacy with an improved survival rate. The addition of liposomal DOX could further improve the endpoint survival rate suggesting that targeting several post-ablation processes with multidrug monotherapies can increase overall ablation efficacy [71].

Although RFA has been approved for treating small-sized unresectable hepatocellular carcinomas (HCCs), treating larger and deeper tumor is always challenging. Thus, a combination approach would be beneficial with intravenous lyso-thermosensitive liposomal doxorubicin (LTLD) along with RFA. This combination therapy showed complete necrosis when RFA is combined with intravenous heat-deployed liposomal doxorubicin therapy in a patient who had liver transplantation 79 days later [72]. Similar study was done by Poon et al. in which LTLD showed increased cure rate of RFA in HCC [73].

Phase I analysis of thermo-liable doxorubicin-liposomal formulation/RFA for hepatic malignancies has been reported by Wood et al. The LTLD formulation was systemically delivered to investigate toxicity, and PK profiles during this combination therapeutic approach. The authors have reported a safe concentration of 50 mg/kg along with RFA as manageable toxicity [74].

The relationship between drug deposition and heating duration was explained by Swenson et al. The ThermoDOX in porcine model showed improved deposition of DOX based on the duration of heating during RFA [75].

Moussa et al. compared the delivery and treatment efficacy of micellar or liposomal drug formulations for the nanodrug-enhanced RFA. The study confirmed that small-sized micellar particles have superior penetration along with enhanced local molecular modulation with RFA. On the other hand, bigger and long-circulating liposomal particles could result in better time-dependent intratumoral drug localization but with reduced tumor growth. Therefore, various factors need to be considered since each formulation has its particular benefits and these should be borne in mind when formulating optimal combination therapies [76].

Recent study by Andriyanov et al. reported the combinatorial therapeutic benefits of PEGylated liposomal DOX/RFA and compared that with single and combined therapy in young and adult mice. Their findings clearly show that age influences tumor advancement and single treatment regimen outcome may significantly be related to tumor microenvironmental changes, and such changes could be surpassed by RFA as it eliminates these differences to enhance the PLD treatment [77].

7.10 Summary and future outlooks

Current RF nanocarriers based on metallic and liposomal-based formulations have shown promising results both in vitro and in vivo. However, more emphasis should be placed on the drug release kinetics and related studies as many of the studies have only focused on how these NPs could be made effective against cancer cells. Metallic NPs combined with novel heat-sensitive nanopolymers or micelles could be another approach to have more controlled release over other systems. More transparent platform is necessary to understand the heating mechanism of RF nano-biomaterials. Novel QDs could be investigated in such a way that is more suitable for theranostic purposes. More research need to be done to fully understand the usability of transition as well as lanthanide QDs along with RF-sensitive nanocarriers and to have better achievements in RF therapy and selectivity of such nanotherapeutics could be improved if one could make them actively target tumor malignancies.

Acknowledgment

This work was supported financially by the Advanced Biomass R&D Center (ABC) of the Global Frontier Project funded by the Ministry of Science and ICT (NRF-2015M3A6A2074238)), and by the Ministry of Science, and ICT (Project No. NRF-2014M3A9E4064580, NRF-2016R1A2B4009619).

References

[1] H.K. Byeon, F.C. Holsinger, R.P. Tufano, H.J. Chung, W.S. Kim, Y.W. Koh, E.C. Choi, Robotic total thyroidectomy with modified radical neck dissection via unilateral retroauricular approach, Ann. Surg. Oncol. 21 (12) (2014) 3872–3875.

[2] H. Nasri, M. Rafieian-Kopaie, World cancer day 2014: "increasing the awareness", Cell J. 16 (3) (2014).

[3] Y. Teng, X. Xie, S. Walker, D.T. White, J.S. Mumm, J.K. Cowell, Evaluating human cancer cell metastasis in zebrafish, BMC Cancer 13 (2013) 453.
[4] W. Li, J. Chang, S. Wang, X. Liu, J. Peng, D. Huang, M. Sun, Z. Chen, W. Zhang, W. Guo, J. Li, miRNA-99b-5p suppresses liver metastasis of colorectal cancer by down-regulating mTOR, Oncotarget 6 (2015) 24448–24462.
[5] C. Xintaropoulou, C. Ward, A. Wise, H. Marston, A. Turnbull, S.P. Langdon, A comparative analysis of inhibitors of the glycolysis pathway in breast and ovarian cancer cell line models, Oncotarget 6 (2015) 25677–25695.
[6] C. Ward, J. Meehan, P. Mullen, C. Supuran, J.M. Dixon, J.S. Thomas, J.Y. Winum, P. Lambin, L. Dubois, N.K. Pavathaneni, E.J. Jarman, L. Renshaw, I. Um, C. Kay, D.J. Harrison, I.H. Kunkler, S.P. Langdon, Evaluation of carbonic anhydrase IX as a therapeutic target for inhibition of breast cancer invasion and metastasis using a series of in vitro breast cancer models, Oncotarget 6 (2015) 24856–24870.
[7] J.W. Spliethoff, E. Tanis, D.J. Evers, B.H. Hendriks, W. Prevoo, T.J. Ruers, Monitoring of tumor radio frequency ablation using derivative spectroscopy, J. Biomed. Opt. 19 (9) (2014) 97004.
[8] X. Fu, Z. Wang, H. Wang, Y.T. Wang, M.W. Jenkins, A.M. Rollins, Fiber-optic catheter-based polarization-sensitive OCT for radio-frequency ablation monitoring, Opt. Lett. 39 (17) (2014) 5066–5069.
[9] Y. Zhong, M. Deng, K. Li, R. Xu, Delayed bronchobiliary fistula and cholangiolithiasis following percutaneous radio frequency ablation for hepatocellular carcinoma, Exp. Biol. Med. (Maywood) 240 (2015) 156–159.
[10] K. Fajkosova, A. Machovcova, M. Onder, K. Fritz, Selective radiofrequency therapy as a non-invasive approach for contactless body contouring and circumferential reduction, J. Drugs Dermatol. 13 (3) (2014) 291–296.
[11] J. Shi, L. Wang, J. Gao, Y. Liu, J. Zhang, R. Ma, R. Liu, Z. Zhang, A fullerene-based multi-functional nanoplatform for cancer theranostic applications, Biomaterials (2014).
[12] F. Peng, Y. Su, X. Ji, Y. Zhong, X. Wei, Y. He, Doxorubicin-loaded silicon nanowires for the treatment of drug-resistant cancer cells, Biomaterials 35 (19) (2014) 5188–5195.
[13] N.S. Rejinold, R. Biswas, G. Chellan, R. Jayakumar, Multifaceted chitin/poly(lactic-co-glycolic) acid composite nanogels, Int. J. Biol. Macromol. 67 (2014) 279–288.
[14] N.S. Rejinold, R. Jayakumar, Y.C. Kim, Radio frequency responsive nano-biomaterials for cancer therapy, J. Control. Release 204 (2015) 85–97.
[15] D.T. Nguyen, W.S. Tzou, L. Zheng, W. Barham, J.L. Schuller, B. Shillinglaw, R.A. Quaife, W.H. Sauer, Enhanced radiofrequency ablation with magnetically directed metallic nanoparticles, Circ. Arrhythm. Electrophysiol. 9 (5) (2016).
[16] N.S. Rejinold, R.G. Thomas, M. Muthiah, H.J. Lee, Y.Y. Jeong, I.K. Park, R. Jayakumar, Breast tumor targetable Fe3O4 embedded thermo-responsive nanoparticles for radiofrequency assisted drug delivery, J. Biomed. Nanotechnol. 12 (1) (2016) 43–55.
[17] H.M. Jhong, C.E. Tornow, C. Kim, S. Verma, J.L. Oberst, P.S. Anderson, A.A. Gewirth, T. Fujigaya, N. Nakashima, P.J.A. Kenis, Gold nanoparticles on polymer-wrapped carbon nanotubes: an efficient and selective catalyst for the electroreduction of CO2, ChemPhysChem 18 (22) (2017) 3274–3279.
[18] U.I. Wu, Y.C. Chuang, W.H. Sheng, H.Y. Sun, Y.T. Jhong, J.Y. Wang, S.C. Chang, J.T. Wang, Y.C. Chen, Use of QuantiFERON-TB Gold In-tube assay in screening for neutralizing anti-interferon-gamma autoantibodies in patients with disseminated nontuberculous mycobacterial infection, Clin. Microbiol. Infect. 24 (2018) 159–165.
[19] E. Seo, J. Ahn, W. Choi, H. Lee, Y.M. Jhon, S. Lee, W. Choi, Far-field control of focusing plasmonic waves through disordered nanoholes, Opt. Lett. 39 (20) (2014) 5838–5841.

[20] M. Jung, J. Koo, J. Park, Y.W. Song, Y.M. Jhon, K. Lee, S. Lee, J.H. Lee, Mode-locked pulse generation from an all-fiberized, Tm-Ho-codoped fiber laser incorporating a graphene oxide-deposited side-polished fiber, Opt. Express 21 (17) (2013) 20062–20072.
[21] J.U. Kim, S.H. Cha, K. Shin, J.Y. Jho, J.C. Lee, Synthesis of gold nanoparticles from gold(I)-alkanethiolate complexes with supramolecular structures through electron beam irradiation in TEM, J. Am. Chem. Soc. 127 (28) (2005) 9962–9963.
[22] M. Ahmed, A.N. Lukyanov, V. Torchilin, H. Tournier, A.N. Schneider, S. N. Goldberg, Combined radiofrequency ablation and adjuvant liposomal chemotherapy: effect of chemotherapeutic agent, nanoparticle size, and circulation time, J. Vasc. Interv. Radiol. 16 (10) (2005) 1365–1371.
[23] J. Lee, D.K. Chatterjee, M.H. Lee, S. Krishnan, Gold nanoparticles in breast cancer treatment: promise and potential pitfalls, Cancer Lett. 347 (2014) 46–53.
[24] L. Karmani, V. Bouchat, C. Bouzin, P. Leveque, D. Labar, A. Bol, G. Deumer, R. Marega, D. Bonifazi, V. Haufroid, C. Michiels, V. Gregoire, O. Feron, S. Lucas, T.V. Borght, B. Gallez, Zr-labelled anti-endoglin antibody-targeted gold nanoparticles for imaging cancer: implications for future cancer therapy, Nanomedicine (London) 9 (2014) 1923–1937.
[25] M.R. Ali, S.R. Panikkanvalappil, M.A. El-Sayed, Enhancing the efficiency of gold nanoparticles treatment of cancer by increasing their rate of endocytosis and cell accumulation using rifampicin, J. Am. Chem. Soc. 136 (2014) 4464–4467.
[26] S.J. Corr, B.T. Cisneros, L. Green, M. Raoof, S.A. Curley, Protocols for assessing radiofrequency interactions with gold nanoparticles and biological systems for non-invasive hyperthermia cancer therapy, J. Vis. Exp. (78) (2013).
[27] S.J. Corr, M. Raoof, Y. Mackeyev, S. Phounsavath, M.A. Cheney, B.T. Cisneros, M. Shur, M. Gozin, P.J. McNally, L.J. Wilson, S.A. Curley, Citrate-capped gold nanoparticle electrophoretic heat production in response to a time-varying radiofrequency electric-field, J. Phys. Chem. C Nanomater. Interfaces 116 (45) (2012) 24380–24389.
[28] H.K. Kim, G.W. Hanson, D.A. Geller, Chemistry. Are gold clusters in RF fields hot or not? Science 340 (6131) (2013) 441–442.
[29] M. Hasegawa, K. Miyata, Y. Abe, T. Ishigami, Radiofrequency heating of metallic dental devices during 3.0 T MRI, Dentomaxillofac. Radiol. 42 (5) (2013) 20120234.
[30] P. Kumari, H. Rickard, P. Majewski, Deposition of silver and gold nanoparticles on surface engineered silica particles and their potential applications, J. Nanosci. Nanotechnol. 12 (10) (2012) 8001–8007.
[31] R.S. McCoy, S. Choi, G. Collins, B.J. Ackerson, C.J. Ackerson, Superatom paramagnetism enables gold nanocluster heating in applied radiofrequency fields, ACS Nano 7 (3) (2013) 2610–2616.
[32] B.H. San, S.H. Moh, K.K. Kim, Investigation of the heating properties of platinum nanoparticles under a radiofrequency current, Int. J. Hyperth. 29 (2) (2013) 99–105.
[33] X. Liu, H.J. Chen, X. Chen, C. Parini, D. Wen, Low frequency heating of gold nanoparticle dispersions for non-invasive thermal therapies, Nanoscale 4 (13) (2012) 3945–3953.
[34] K. Miyata, M. Hasegawa, Y. Abe, T. Tabuchi, T. Namiki, T. Ishigami, Radiofrequency heating and magnetically induced displacement of dental magnetic attachments during 3.0 T MRI, Dentomaxillofac. Radiol. 41 (8) (2012) 668–674.
[35] D. Li, Y.S. Jung, S. Tan, H.K. Kim, E. Chory, D.A. Geller, Negligible absorption of radiofrequency radiation by colloidal gold nanoparticles, J. Colloid Interface Sci. 358 (1) (2011) 47–53.
[36] C. Schmidt, The Kanzius machine: a new cancer treatment idea from an unexpected source, J. Natl. Cancer Inst. 100 (14) (2008) 985–986.
[37] Y. Xu, M. Mahmood, Z. Li, E. Dervishi, S. Trigwell, V.P. Zharov, N. Ali, V. Saini, A.R. Biris, D. Lupu, D. Boldor, A.S. Biris, Cobalt nanoparticles coated with graphitic

shells as localized radio frequency absorbers for cancer therapy, Nanotechnology 19 (43) (2008) 435102.
[38] M. Raoof, C. Zhu, W.D. Kaluarachchi, S.A. Curley, Luciferase-based protein denaturation assay for quantification of radiofrequency field-induced targeted hyperthermia: developing an intracellular thermometer, Int. J. Hyperth. 28 (3) (2012) 202–209.
[39] X. Huang, P.K. Jain, I.H. El-Sayed, M.A. El-Sayed, Determination of the minimum temperature required for selective photothermal destruction of cancer cells with the use of immunotargeted gold nanoparticles, Photochem. Photobiol. 82 (2) (2006) 412–417.
[40] N.S. Rejinold, R. Ranjusha, A. Balakrishnan, N. Mohammed, R. Jayakumar, Gold–chitin–manganese dioxide ternary composite nanogels for radiofrequency assisted cancer therapy, RSC Adv. 4 (11) (2014) 5819–5825.
[41] M. Banobre-Lopez, A. Teijeiro, J. Rivas, Magnetic nanoparticle-based hyperthermia for cancer treatment, Rep. Pract. Oncol. Radiother. 18 (6) (2013) 397–400.
[42] M.C. Fastame, P.K. Hitchcott, M.P. Penna, Do self-referent metacognition and residential context predict depressive symptoms across late-life span? A developmental study in an Italian sample, Aging Ment. Health (2014) 1–7.
[43] F.E. Demircioglu, P. Burkhardt, D. Fasshauer, The SM protein Sly1 accelerates assembly of the ER-Golgi SNARE complex, Proc. Natl. Acad. Sci. U. S. A. 111 (38) (2014) 13828–13833.
[44] M.F. Spadea, A. Fassi, P. Zaffino, M. Riboldi, G. Baroni, N. Depauw, J. Seco, Contrast-enhanced proton radiography for patient set-up by using X-Ray CT prior knowledge, Int. J. Radiat. Oncol. Biol. Phys. 90 (3) (2014) 628–636.
[45] A.E. Fasang, M. Raab, Beyond transmission: intergenerational patterns of family formation among middle-class American families, Demography 51 (2014) 1703–1728.
[46] J.T. Kowallick, A.A. Joseph, C. Unterberg-Buchwald, M. Fasshauer, K. van Wijk, K.D. Merboldt, D. Voit, J. Frahm, J. Lotz, J.M. Sohns, Real-time phase-contrast flow MRI of the ascending aorta and superior vena cava as a function of intrathoracic pressure (Valsalva manoeuvre), Br. J. Radiol. 87 (1042) (2014) 20140401.
[47] E. Carenza, V. Barcelo, A. Morancho, L. Levander, C. Boada, A. Laromaine, A. Roig, J. Montaner, A. Rosell, In vitro angiogenic performance and in vivo brain targeting of magnetized endothelial progenitor cells for neurorepair therapies, Nanomedicine 10 (1) (2014) 225–234.
[48] M. Heiden, E. Walker, E. Nauman, L. Stanciu, Evolution of novel bioresorbable iron-manganese implant surfaces and their degradation behaviors in vitro, J. Biomed. Mater. Res. A 103 (2014) 185–193.
[49] I. Akartuna, E. Tervoort, A.R. Studart, L.J. Gauckler, General route for the assembly of functional inorganic capsules, Langmuir 25 (21) (2009) 12419–12424.
[50] M.N. Chandraprabha, K.N.L. Lokesh, K.P. Ramesh, R.B. Rashmi, R.G. Roshna, S.N. Vijai, Evaluation of hyperthermia effect of magnetic iron oxide nanoparticles coated with different polymers, J. Chem. Eng. Res. 2 (2014) 77–84.
[51] N.S. Rejinold, K.P. Chennazhi, R. Jayakumar, One pot green synthesis of iron oxide nanoparticles by O-carboxymethyl chitosan assisted hydrothermal method, J. Chitin Chitosan Sci. 1 (1) (2013) 76–85.
[52] M. Thikra, Z. Yongbin, W. Fumiya, K. Alokita, P.A. Madhu, L. Reginald, H.M. Keith, X. Yang, S.B. Alexandru, S. Biris, Iron oxide nanoparticle-based radiofrequency thermotherapy for human breast adenocarcinoma cancer cells, Biomater. Sci. 1 (2013) 870–880.
[53] K.H. Bae, M. Park, M.J. Do, N. Lee, J.H. Ryu, G.W. Kim, C. Kim, T.G. Park, T. Hyeon, Chitosan oligosaccharide-stabilized ferrimagnetic iron oxide nanocubes for magnetically modulated cancer hyperthermia, ACS Nano 6 (6) (2012) 5266–5273.

[54] P.M. Peiris, L. Bauer, R. Toy, E. Tran, J. Pansky, E. Doolittle, E. Schmidt, E. Hayden, A. Mayer, R.A. Keri, M.A. Griswold, E. Karathanasis, Enhanced delivery of chemotherapy to tumors using a multicomponent nanochain with radio-frequency-tunable drug release, ACS Nano 6 (5) (2012) 4157–4168.
[55] A. Karmakar, Y. Xu, M.W. Mahmood, Y. Zhang, L.M. Saeed, T. Mustafa, S. Ali, A.R. Biris, A.S. Biris, Radio-frequency induced in vitro thermal ablation of cancer cells by EGF functionalized carbon-coated magnetic nanoparticles, J. Mater. Chem. 21 (34) (2011) 12761–12769.
[56] Y. Xu, A. Karmakar, W.E. Heberlein, T. Mustafa, A.R. Biris, A.S. Biris, Multifunctional magnetic nanoparticles for synergistic enhancement of cancer treatment by combinatorial radio frequency thermolysis and drug delivery, Adv. Healthc. Mater. 1 (4) (2012) 493–501.
[57] S.D. Kong, W. Zhang, J.H. Lee, K. Brammer, R. Lal, M. Karin, S. Jin, Magnetically vectored nanocapsules for tumor penetration and remotely switchable on-demand drug release, Nano Lett. 10 (12) (2010) 5088–5092.
[58] E.S. Glazer, S.A. Curley, Radiofrequency field-induced thermal cytotoxicity in cancer cells treated with fluorescent nanoparticles, Cancer 116 (13) (2010) 3285–3293.
[59] X. Yang, K. Alokita, W. Daoyuan, W.M. Meena, W. Fumiya, Z. Yongbin, F. Ashley, F. Phillip, L. Zhongrui, K. Ganesh, Syed Ali, R.B. Alexandru, S.B. Alexandru, Multifunctional Fe_3O_4 cored magnetic-quantum dot fluorescent nanocomposites for RF nanohyperthermia of cancer cells, J. Phys. Chem. C 114 (11) (2010) 5020–5026.
[60] J. Bhardwaj, S. Devarakonda, S. Kumar, J. Jang, Development of a paper-based electrochemical immunosensor using an antibody-single walled carbon nanotubes bioconjugate modified electrode for label-free detection of foodborne pathogens, Sensor. Actuat. B-Chem. 253 (2017) 115–123.
[61] H.H. Byeon, W.C. Lee, W. Kim, S.K. Kim, W. Kim, H. Yi, Bio-fabrication of nanomesh channels of single-walled carbon nanotubes for locally gated field-effect transistors, Nanotechnology 28 (2) (2017).
[62] A. Gasnier, J.M. Gonzalez-Dominguez, A. Anson-Casaos, J. Hernandez-Ferrer, M.L. Pedano, M.D. Rubianes, M.T. Martinez, G. Rivas, Single-wall carbon nanotubes covalently functionalized with polylysine: synthesis, characterization and analytical applications for the development of electrochemical (bio)sensors, Electroanalysis 26 (8) (2014) 1676–1683.
[63] J. Lee, A. Efremov, R.G. Son, S.P. Pack, H.W. Lee, K. Kim, K.H. Kwon, Ammonia-based plasma treatment of single-walled carbon nanotube thin films for bio-immobilization, Carbon 105 (2016) 430–437.
[64] C.J. Gannon, P. Cherukuri, B.I. Yakobson, L. Cognet, J.S. Kanzius, C. Kittrell, R.B. Weisman, M. Pasquali, H.K. Schmidt, R.E. Smalley, S.A. Curley, Carbon nanotube-enhanced thermal destruction of cancer cells in a noninvasive radiofrequency field, Cancer 110 (12) (2007) 2654–2665.
[65] A. Sasidharan, A.J. Sivaram, A.P. Retnakumari, P. Chandran, G.L. Malarvizhi, S. Nair, M. Koyakutty, Radiofrequency ablation of drug-resistant cancer cells using molecularly targeted carboxyl-functionalized biodegradable graphene, Adv. Healthc. Mater. 4 (5) (2015) 679–684.
[66] D. Bijukumar, C.M. Girish, A. Sasidharan, S. Nair, M. Koyakutty, Transferrin-conjugated biodegradable graphene for targeted radiofrequency ablation of hepatocellular carcinoma, ACS Biomater. Sci. Eng. 1 (12) (2015) 1211–1219.
[67] V.H. Somasundaram, R. Pillai, G. Malarvizhi, A. Ashokan, S. Gowd, R. Peethambaran, S. Palaniswamy, A.K.K. Unni, S. Nair, M. Koyakutty, Biodegradable radiofrequency responsive nanoparticles for augmented thermal ablation combined with triggered drug release in liver tumors, ACS Biomater. Sci. Eng. 2 (5) (2016) 768–779.

[68] M.M.P. Kuijten, M.H. Degeling, J.W. Chen, G. Wojtkiewicz, P. Waterman, R. Weissleder, J. Azzi, K. Nicolay, B.A. Tannous, Multimodal targeted high relaxivity thermosensitive liposome for in vivo imaging, Sci. Rep. 5 (2015) 17220.
[69] A.V. Andriyanov, E. Koren, Y. Barenholz, S.N. Goldberg, Therapeutic efficacy of combining PEGylated liposomal doxorubicin and radiofrequency (RF) ablation: comparison between slow-drug-releasing, non-thermosensitive and fast-drug-releasing, thermosensitive nano-liposomes, PLoS One 9 (5) (2014).
[70] W. Yang, M. Ahmed, B. Tasawwar, T. Levchenko, R.R. Sawant, M. Collins, S. Signoretti, V. Torchilin, S.N. Goldberg, Radiofrequency ablation combined with liposomal quercetin to increase tumour destruction by modulation of heat shock protein production in a small animal model, Int. J. Hyperth. 27 (6) (2011) 527–538.
[71] W. Yang, M. Ahmed, B. Tasawwar, T. Levchenko, R.R. Sawant, V. Torchilin, S.N. Goldberg, Combination radiofrequency (RF) ablation and IV liposomal heat shock protein suppression: reduced tumor growth and increased animal endpoint survival in a small animal tumor model, J. Control. Release 160 (2) (2012) 239–244.
[72] C.W. Hong, S.K. Libutti, B.J. Wood, Liposomal doxorubicin plus radiofrequency ablation for complete necrosis of a hepatocellular carcinoma, Curr. Oncol. 20 (3) (2013) e274–e277.
[73] R.T. Poon, N. Borys, Lyso-thermosensitive liposomal doxorubicin: an adjuvant to increase the cure rate of radiofrequency ablation in liver cancer, Future Oncol. 7 (8) (2011) 937–945.
[74] B.J. Wood, R.T. Poon, J.K. Locklin, M.R. Dreher, K.K. Ng, M. Eugeni, G. Seidel, S. Dromi, Z. Neeman, M. Kolf, C.D. Black, R. Prabhakar, S.K. Libutti, Phase I study of heat-deployed liposomal doxorubicin during radiofrequency ablation for hepatic malignancies, J. Vasc. Interv. Radiol. 23 (2) (2012) 248–255 e7.
[75] C.E. Swenson, D. Haemmerich, D.H. Maul, B. Knox, N. Ehrhart, R.A. Reed, Increased duration of heating boosts local drug deposition during radiofrequency ablation in combination with thermally sensitive liposomes (ThermoDox) in a porcine model, PLoS One 10 (10) (2015) e0139752.
[76] M. Moussa, S.N. Goldberg, G. Kumar, R.R. Sawant, T. Levchenko, V.P. Torchilin, M. Ahmed, Nanodrug-enhanced radiofrequency tumor ablation: effect of micellar or liposomal carrier on drug delivery and treatment efficacy, PLoS One 9 (8) (2014) e102727.
[77] A.V. Andriyanov, E. Portnoy, E. Koren, S. Inesa, S. Eyal, S.N. Goldberg, Y. Barenholz, Therapeutic efficacy of combined PEGylated liposomal doxorubicin and radiofrequency ablation: Comparing single and combined therapy in young and old mice, J. Control. Release 257 (2017) 2–9.

CHAPTER 8

Targeting peptide-modified polymeric nanoparticles for cardiac-specific drug delivery applications

Muthunarayanan Muthiah, Jason McCarthy
Masonic Medical Research Laboratory, Utica, NY, United States

Contents

8.1 Introduction

Nanoparticles are the substances, which are in the size range of 10–100 nm. Nanoparticles posses larger surface area and unique optical, electrical, and mechanical properties when compared to the bulk. Surface modification of the nanoparticles helps in improving the properties depending on the application. The use of nanomaterials for the delivery of the drug or gene to the biological system in vitro and in vivo is the discipline of nanomedicine. Biocompatibility is the major issue when the nanoparticles are used for the biological applications. The modification of the nanoparticle surface with different functional group and polymers like polyethylene glycol helps in alleviating the toxicity in vitro and in vivo. Specific targeting of the

Biomimetic Nanoengineered Materials for Advanced Drug Delivery
https://doi.org/10.1016/B978-0-12-814944-7.00008-4

nanoparticle to the region of interest is necessary to reduce the off-target effects. Targeting can be achieved by attaching the moieties like antibody, peptide, aptamer, carbohydrates, etc., to the nanoparticle, which will guide the particle to the specific area of interest. This method of targeting is called the active targeting [1].

Passive targeting utilizes the leaky vasculature in the inflamed tissue or diseased area for the entry of the nanoparticle into the cells and tissues in the affected region of the body. After reaching the region of interest the nanoparticles have to release the loaded drug or gene. Chemical drugs are generally hydrophobic in nature, so the hydrophilic materials are used for loading the drug into the core with the help of polymeric or lipid-constituted micelle nanocarriers. Therapeutic gene can be complexed with the cationic nanoparticles and can be injected into the body. Once the cells take up the nanoparticle with the gene, the endosomal pH is altered due the increase in the positive charge and the gene will be released in the cytoplasm. In the case of the plasmid DNA it has to reach the nucleus for the expression, whereas in the case of microRA and siRNA, the processing enzymes at the site of release will act upon them [2–5].

8.2 Cardiovascular disease

Cardiovascular diseases are the leading cause of death in the developed and developing countries. Major damage occurs due to the cardiac ischemia, commonly known as a heart attack, resulting from the interruption of blood supply to a part of the heart, causing heart cells to die. Current therapeutic intervention includes coronary angioplasty, stent implantation, coronary artery by-pass grafting. After stent placement, local injury promotes the vascular smooth cell proliferation resulting in restenosis. Drug-loaded stents suppress the proliferation of the smooth muscle cells. Defects in the drug-eluting stents are late stent thrombosis, inflammation, arterial healing. Vascular smooth cell-targeted small molecules are useful for delivering antiproliferative drug or gene loaded with nanocarrier. Endothelial cells and macrophages in the cardiovascular system can be targeted with the help of the targeting ligands to deliver the therapeutic drug (Fig. 8.1).

8.3 CTP peptide-mediated targeting to cardiomyocytes

Zahid et al. have identified a cardiac-specific protein transduction domain by in vivo biopanning using an M13 phage peptide display library in mice [6].

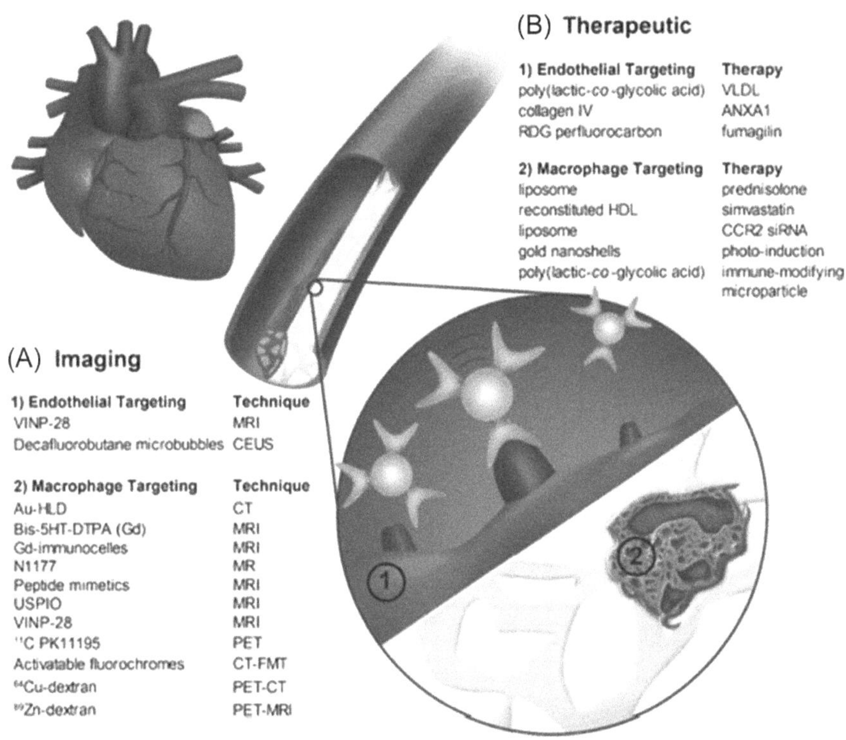

Fig. 8.1 Nanotechnology approaches targeting endothelial and macrophage populations in atherosclerosis. Within the atheroma nanotechnology has been applied for both imaging and therapeutic delivery to the endothelial cell (1) and macrophage (2). (A) Imaging applications in atherosclerosis include both nanoparticle description and imaging technique. (B) Therapies applied with both description of nanoparticle design and therapeutic payload. *Figure reproduced after getting the permission from springer (License Number 4478820295179)*

With this CTP peptide they were able to deliver fluorescently labeled Streptavidin, a, 60-kDa complex, to cardiac cells in vivo without transduction of liver, spleen, skeletal muscle, or brain, with minimal uptake by lung and glomerular capillaries.

Using these CTP peptides Chien et al. have delivered the cre peptide specifically to the cardiac cells and assessed the cre protein recombinase activity. Initially they treated this His-cre-CTP peptide with the cells that constitutively express membrane-bound tdTomato fluorescent protein (mT). In the presence of HIS-cre-CTP, the cells undergo excision of the open reading frame of mT, resulting in the expression of membrane-bound green fluorescent protein.

In order to deliver cre CTP in a prolonged manner effectively they have combined the protein PF127 polymer and delivered it efficiently in muscles. They have shown that protein mixing with PF-127 hydrogel facilitates injection and uptake of enzymatically active protein into skeletal muscle in living animals. Together with ultrasound guidance, the method has been used to deliver enzymatically active protein directly to cardiac muscle.

Avula et al. have conjugated these CTP peptides to nanoparticles for specific delivery of photosensitizer to cardiomyocytes [2]. They have tested this material for photodynamic therapy in vitro. They have demonstrated that CTP-nanoparticle photosensitizer has the unique capability to specifically attach to myocytes and not to fibroblasts and to induce cell-specific death upon local laser light delivery, followed by local release of ROS. In the illuminated areas decreased number of viable myocytes were observed, while the number of healthy fibroblasts was not affected by the treatment.

By this work they provided the proof-of-principle for a novel targeted cardiac ablation technology that could possibly achieve cell and spatial specificity. Due to the spatial specificity, photodynamic ablation enables local confinement of the therapeutic effect, minimizing the adverse damage to adjacent nontargeted cells and tissues.

8.4 Identification and application of PCM peptide for cardiac targeting

McGuire et al. have identified PCM1 (WLSEAGPVVTVRALRGTGSW) peptide for targeting cardiomyocytes [7]. The PCM1 peptide was selected from their library of 20-mer peptides, displayed on filamentous phage, by reiterative rounds of biopanning on primary cardiomyocytes. Cardiomyocytes have exhibited high selectivity for this PCM1 peptide compared to other peptides in their library and the PCM1 peptide has high specificity for cardiomyocytes compared to other cell types. PCMP 1 was selected based on in vitro screening and they have demonstrated that cardiomyocytes accumulated PCM1 phage approximately 1000-fold over RC phage in vivo. PCM1 phage accumulated in heart tissue at 5- to 180-fold higher levels than other tissues in vivo. They have confirmed the preferential accumulation of the PCM phage in heart and the distribution in other tissues by using real-time PCR detection of the phage DNA.

Hye et al. have used these PCM peptide for conjugation with the bioreducible poly(cystamine bisacrylamide-diaminohexane, CBA-DAH). As previously described PCM was an isolated phage that displays a 20-amino acid peptide (WLSEAGPVVTVRALRGTGSW) and binds primary cardiomyocytes 180 times more avidly than the other control phages. They have introduced the cysteine amino acid to the PCM and conjugated the cysteine-terminated PCM to an activated poly(CBA-DAH) using an SM(PEG)2 cross-linker. This carrier was used to deliver Fas siRNA for downregulating the Fas gene expression and to inhibit cardiomyocyte apoptosis. From their studies they were able to show that a bioreducible polymer conjugated with cysteine-modified PCM residues has the ability to differentiate between cardiomyocyte and non-cardiomyocyte cell types effectively.

8.5 AT1 peptide-mediated targeting of the cardiovascular system

Dvir et al. have utilized the AT1 receptor specific ligand for targeting the heart after myocardial infarction (MI) [8]. Liposomes were conjugated to the peptide with these sequence (NH2-GGGGYDRVTIHPF-NH2) for targeting the heart. The nanoparticles were able to specifically target cardiac cells in vitro, and in the infarcted heart after intravenous injection in vivo. They delivered these nanoformulations through intravenous injection, which avoids the need for cardiac catheterization or other surgical procedures.

Based on their work they have demonstrated that nanoparticles were targeted to the injured heart and not to the healthy, suggesting this a valuable method for the delivery of drugs after MI. Peptide ligand-mediated targeting further enhances the delivery to the injured myocardium, which will be of greater advantage when considering the whole system.

8.6 Vascular smooth muscle-targeting ligands

8.6.1 Restenotic vascular smooth muscle cell-targeting peptide

Ingrid et al. have used phage display technique to identify the peptide, which is targeted to proliferating vascular smooth muscle cells [9]. They have identified two consensus 5G6 (CNIWGVVLSWIGVFPEC) and 5E5 (CESLWGGLMWTIGLSDC) after in vitro and in vivo biopanning. IGR motif was found to recur in the in vivo clones. The consensus phages 5G6 and 5E5 bind selectively to VSMCs compared to other cell types. Preferential binding of the peptides to the proliferating VSMCs compared to growth-arrested VSMCs was observed by the authors in this study. Proliferating smooth muscle cells are the dominant cell types in restenotic lesions, so drug or gene, which suppress the VSMCs, can be loaded to these peptide-bound nanocarrier for the restenosis therapy. In future, using these peptides, the specific receptor for the VSMC can be discovered, since no cell surface receptor specific for VSMC has yet been identified yet. Once the VSMC-specific receptor is identified, the peptide and the receptor can be used for drug or gene targeting, as well as imaging studies.

8.6.2 RNA-aptamer targeting the smooth muscle cell

Aptamers are single-stranded DNA or RNA molecules. Aptamers can be in different shapes as result of helics and loops formation. They are extremely versatile and bind targets with high selectivity and specificity.

William et al. have used an RNA aptamer (Apt 14) to preferentially target the VSMCs as compared to ECs and other myocytes [10]. Apt 14 is a inhibitor of the vascular smooth muscle cell migration by the interruption of PI3K/Akt-dependent signaling pathway. In the in vivo studies they have observed that the APT 14 inhibits the neointimal formation in vascular injured mice model. The level of neoitimal reduction due to APT 14 was comparable to the drug paclitaxel. APT14 reacted with murine and human VSMC without eliciting the immune response in the respective models. Drug for the VSMC proliferation reduction results in delayed endothelialization and late stent thrombosis, but the aptamer-mediated cell-targeting therapies will reduce the VSMC migration while allowing the endothelialization. This will aid in enhanced vascular healing and the prevention of restenosis.

8.7 Targeted macrophage ablation in inflammatory atherosclerosis

Jason et al. developed a theranostic nanoagent for targeting the macrophage in the inflammatory atherosclerotic lesions [11]. Magnetic nanoparticles (CLIO—cross-linked iron oxide) were modified with near-infrared flourophore to develop the chlorin-based photosensitizer. They have investigated the macrophage-specific accumulation in vitro and in vivo in a murine model of atherosclerosis. Intravenous administration results in the localization of the nanoagent within macrophage-rich atherosclerotic lesions that can be imaged by intravital fluorescence microscopy. Irradiation of atheroma with 650-nm light activates the therapeutic component and results in the eradication of inflammatory macrophages, which may induce

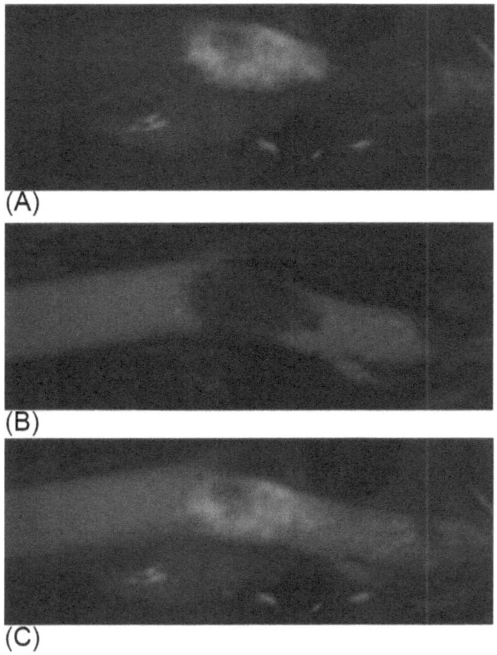

Fig. 8.2 In vivo localization of the nanoagent, CLIO-THPC, to carotid atheroma, as determined by intravital fluorescence microscopy. (A) Fluorescence image in the AF750 channel demonstrating particle uptake by a carotid plaque. (B) Fluorescence angiogram utilizing fluorescein-labeled dextran to outline the vasculature. (C) Merged image of the two fluorescence channels.

lesion stabilization. This nanoagent is toxic only upon light activation whereas the small molecule drugs have a lot of off-target effects. Theranostic nanoagents will help in therapy and diagnosis; this nanoagent has proved its mettle as a theranostic material by helping in imaging atherosclerosis and in their treatment by photodynamic therapy (Fig. 8.2).

8.8 Conclusion

Targeting therapeutic drug or gene to specific organ of interest has been a hot topic of research in the recent decade. Cardiovascular targeting has gained the interest of researchers, clinicians and pharmaceutical companies. MI and stroke are among the main causes of mortality in United States and other developing countries. Molecular imaging for understanding the disease progression with molecular imaging agents will be of great aid for the treatment. These new targeted imaging agents can help in providing specific individual-based medical treatment, monitor the progress of the interventions, and detection of plaque and its progression in vivo.

References

[1] M. Muthiah, I.K. Park, C.S. Cho, Surface modification of iron oxide nanoparticles by biocompatible polymers for tissue imaging and targeting, Biotechnol Adv 31 (8) (2013) 1224–1236.
[2] U.M. Avula, H.K. Yoon, C.H. Lee, K. Kaur, R.J. Ramirez, Y. Takemoto, S.R. Ennis, F. Morady, T. Herron, O. Berenfeld, R. Kopelman, J. Kalifa, Cell-selective arrhythmia ablation for photomodulation of heart rhythm, Sci Transl Med 7 (311) (2015). 311ra172.
[3] M. Muthiah, H.L. Che, S. Kalash, J. Jo, S.Y. Choi, W.J. Kim, C.S. Cho, J.Y. Lee, I.K. Park, Formulation of glutathione responsive anti-proliferative nanoparticles from thiolated Akt1 siRNA and disulfide-crosslinked PEI for efficient anti-cancer gene therapy, Colloids Surf B Biointerfaces 126 (2015) 322–327.
[4] M. Muthiah, M.A. Islam, H.J. Lee, M.J. Moon, C.S. Cho, I.K. Park, MicroRNA delivery with osmotic polysorbitol-based transporter suppresses breast cancer cell proliferation, Int J Biol Macromol 72 (2015) 1237–1243.
[5] M. Muthiah, S.H. Park, M. Nurunnabi, J. Lee, Y.K. Lee, H. Park, B.I. Lee, J.J. Min, I.K. Park, Intracellular delivery and activation of the genetically encoded photosensitizer Killer Red by quantum dots encapsulated in polymeric micelles, Colloids Surf B Biointerfaces 116 (2014) 284–294.
[6] M. Zahid, K.S. Feldman, G. Garcia-Borrero, T.N. Feinstein, N. Pogodzinski, X. Xu, R. Yurko, M. Czachowski, Y.L. Wu, N.S. Mason, C.W. Lo, Cardiac targeting peptide, a novel cardiac vector: studies in bio-distribution, imaging application, and mechanism of transduction, Biomolecules 8 (4) (2018).
[7] M.J. McGuire, K.N. Samli, S.A. Johnston, K.C. Brown, In vitro selection of a peptide with high selectivity for cardiomyocytes in vivo, J Mol Biol 342 (1) (2004) 171–182.
[8] T. Dvir, M. Bauer, A. Schroeder, J.H. Tsui, D.G. Anderson, R. Langer, R. Liao, D. S. Kohane, Nanoparticles targeting the infarcted heart, Nano Lett 11 (10) (2011) 4411–4414.
[9] I.N. Michon, A.D. Hauer, J.H. von der Thusen, T.J. Molenaar, T.J. van Berkel, E.A. Biessen, J. Kuiper, Targeting of peptides to restenotic vascular smooth muscle cells using phage display in vitro and in vivo, Biochim Biophys Acta 1591 (1-3) (2002) 87–97.
[10] W.H. Thiel, C.L. Esposito, D.D. Dickey, J.P. Dassie, M.E. Long, J. Adam, J. Streeter, B. Schickling, M. Takapoo, K.S. Flenker, J. Klesney-Tait, V. Franciscis, F.J. Miller Jr., P.H. Giangrande, Smooth muscle cell-targeted RNA aptamer inhibits neointimal formation, Mol Ther 24 (4) (2016) 779–787.
[11] J.R. McCarthy, E. Korngold, R. Weissleder, F.A. Jaffer, A light-activated theranostic nanoagent for targeted macrophage ablation in inflammatory atherosclerosis, Small 6 (18) (2010) 2041–2049.

CHAPTER 9

Nanogel-based active drug delivery

Johnson V. John
Department of Chemical and Materials Engineering, University of Alberta, Edmonton, AB, Canada

Contents

9.1 Introduction

Standing next to the nanocarriers, nanogels (NGs) have shown their promising potential for the controlled accommodation of payloads and their delivery into pathological sites amid other nanocarriers including micelles, liposomes, and polymerosome [1]. NGs are a kind of cross-linked hydrogel system with size ranging from 20 to 200 nm with high encapsulation efficiency and protection of active agents from degradation during circulation, which make them ideal candidates in nanomedicine research stream [2, 3]. Generally, three-dimensional (3D) NGs formed by either physical or chemical cross-linking of various macromolecular chains are able to swell or hold water without changing its 3D structure. Smart NGs are able to formulate by responding to various stimuli like pH, temperature, redox, enzyme, etc., by incorporating the respective macromolecule moiety as a precursor during the NG formulation [4–6]. The above-mentioned features of NGs help to surface as a promising material in nanomedicine, especially anticancer therapy.

Biomimetic Nanoengineered Materials for Advanced Drug Delivery
https://doi.org/10.1016/B978-0-12-814944-7.00009-6

9.2 Synthesis of NGs

NGs are classified into two groups: (i) physically cross-linked NGs and (ii) chemically cross-linked NGs. Physically cross-linked NGs are generally formed by non-covalent weak forces like van der Waals forces, hydrophobic, electrostatic interactions, and hydrogen bonding. Moreover, the strength and reliability of the physically cross-linked NGs always depend on composition and cross-linking density of the monomer/polymers during formulation [7, 8]. Particularly, ionic charge-induced NG formation will be performed between the oppositely charged polymeric chains results in few minutes during the formulation [9]. But chemically cross-linked NGs prepared by using various cross-linking reactions undergo composed polymer/ monomer units. The covalent cross-linking strategy has been influenced the properties of NGs for the controlled loading of therapeutics cargo into NGs matrices. Especially the strength, swellability, porosity, and morphology of the NGs always depend on the controlled delivery of therapeutic cargo into the predetermined sites. Various kinds of monomer/cross-linkers including *N*,*N*′-methylenebisacrylamide, divinylbenzene, diallyl phthalate, PEG diacrylate and redox-responsive diacrylated cystamine, and acetal bearing cross-linker have been widely used to construct chemically cross-linked NGs over the past few decades [10].

9.3 Advantages of NGs as a drug carrier

The intracellular delivery of therapeutic payload is challenging to cross the physiological barriers (enzymatic and chemical degradation) and cellular membrane to reach a specific site. Undoubtedly, NGs are one of the ideal candidates among other nanocarriers including the liposomes, micelles, and polymersomes, etc., since it can deliver payloads via simple administration protocols like oral, pulmonary, nasal, parenteral, intraocular, and topical routes into blood stream [2–4]. Moreover, the hydrophilicity of NGs offers some incredible features including biocompatibility and high-loading capacity for hydrophilic biotherapeutics, and their 3D network protects the encapsulated molecules against degradation because enzymes cannot penetrate into the particles as compared to other above-mentioned carriers [7–10].

(i) Size and surface properties of the NGs can be tuned by feed ratio of the components (during preparation), then the surface properties including permeation capability, escaping the renal clearance by phagocytic cells, and uptake by the reticuloendothelial system, will be enhanced. Moreover, it must be smart enough to penetrate tissues through paracellular or transcellular pathways.

(ii) NGs can be prepared via very simple and easy procedures such as cross-linking, radical polymerization, click chemistry. Moreover, the involvement of organic solvents is negligible as compared to the synthesis of other nanocarriers. Importantly, NGs will be interesting for clinical markets due to the "green" approaches and large scale synthesis for NGs manufacturing.

(iii) The swelling or deswelling nature (water medium) of the NGs is most beneficial characteristics as compared to other nanocarriers for the controlled delivery of payloads in physiological conditions.

(iv) Active drug-loading capacity is relatively high as compared to other nanocarriers due to the cross-linking capacity, ionic interaction, weak bonds in the NG 3D network.

(v) NGs are prepared to transport therapeutic cargo into the specified site and it will be more accurate by decorating the specific ligands in NG surface to void accumulation of payloads in normal tissues.

(vi) The incorporation of stimuli-responsive moieties into NGs can be more effective for the controlled delivery of payloads because the drug-loaded NGs are stable during circulation, but after reaching the targeted site the payload should be released eventually in response to the appropriate intracellular or extracellular milieu.

(vii) Most of the NGs exhibit slower rates of dissociation, better stability over the surfactant micelles, lower critical micelle concentrations, and, most importantly, longer retention of loaded drugs.

The aforementioned features of the NGs show its prominence in drug delivery platform and the stimuli-responsive NG feature has been described in the following part of this chapter. The types and significance of stimuli-responsive NGs for the controlled delivery of therapeutic cargo in anticancer therapy are briefly described.

9.4 Mechanism of active drug delivery from NGs

For the past few decades, stimuli-responsive nanocarriers have been receiving more attention for the controlled delivery of therapeutics in anticancer therapy [11, 12]. Among them, stimuli-responsive NGs have been widely used for the controlled delivery of the drug, gene or other anticancer molecules to the targeted site. Interestingly, the colloidal and morphological

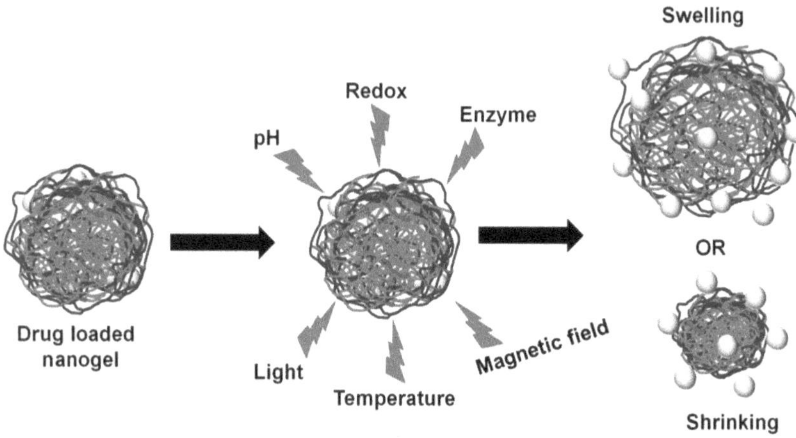

Fig. 9.1 Stimuli-responsive drug release from nanogel.

features of these stimuli-sensitive NGs will impeccably accommodate therapeutic payloads and release them by changing their volume (swelling/shrinking) due to the exogenous or endogenous stimuli of the media/environment (Fig. 9.1) [1, 2, 4].

9.4.1 Exogenous stimuli-responsive NGs

In this section, we discuss the exogenous stimuli-responsive nanocarriers such as externally applied stimulus, like temperature, magnetic fields, and light.

Temperature-responsive NGs: Generally, temperature-responsive NGs respond to temperature (increase or decrease), which leads to the sustained delivery of therapeutic payloads into the specific site. The temperature which causes rapid changes in the particle size of NGs (swelling or shrinking) is usually called volume phase transition temperature (VPTT) [13, 14]. Ideally, these temperature-responsive polymers demonstrate their thermal response in the biologically important range of 10–40°C. However, the swelling of NGs is more efficient for controlled and sustained delivery of therapeutics in the specific site as compared to shrinking NGs, since it could not release therapeutics completely due to shrinking. While designing a temperature-responsive NGs, the VPTT of the NGs will be a bit higher than physiological temperature because the temperature is usually elevated in inflamed tissues where therapeutics are being delivered, and hyperthermia can also be induced locally by a number of external heating techniques.

Poly(*N*-isopropylacrylamide) (PNIPAM) and poly(*N*-vinylcaprolactam) (PNVCL) are the common temperature-responsive polymers which show a lower critical solution temperature (LCST) close to the physiological conditions [13]. Therefore, researchers have been trying to develop ideal temperature-responsive NGs by using PNIPAM and PVCL. PNIPAM-incorporated NGs has been generally prepared by emulsion polymerization with the combination of various monomers (acrylic acid methacrylic acid, vinyl acetic acid, acrylamide, maleic acid, fumaric acid, allyl-acetic acid, acrylonitrile, and aminoethylmethacrylate hydrochloride). In 2009, Kawano et al. prepared hybrid NGs (a single gold nanorod encapsulated in a PNIPAM NGs) by colloid-template polymerization and silica etching, the phase transition of the NG particle has been controlled by applying NIR (near-infrared) irradiation into local targeted site [15]. Interestingly, Calderon's group has reported few articles on PNIPAM including semi-interpenetrating temperature-responsive NGs for the active delivery of drugs through the hair follicles (Fig. 9.2) [16, 17]. Apart from the PNIPAM, PNVCL is a good choice to fabricate temperature-responsive NGs since it shows more biocompatibility and less hydrolysis as compared to PNIPAM. Gao et al. reported first PVCL-based NGs, later on, significant research efforts has been devoted to the development of PNVCL-based NGs [18]. Recently, Kharlampieva's group has developed PNVCL NG containing multilayer film for the tropical delivery of therapeutics [19].

Photo-/light-responsive NGs: Photo-responsive NGs can be fabricated from monomers bearing photoactive groups that alter bonding or conformation in response to light exposure. Common light-responsive functional groups like spirobenzopyran, cinnamonyl, and azobenzene show their phase transition upon light exposure. Unfortunately, drug delivery based on this

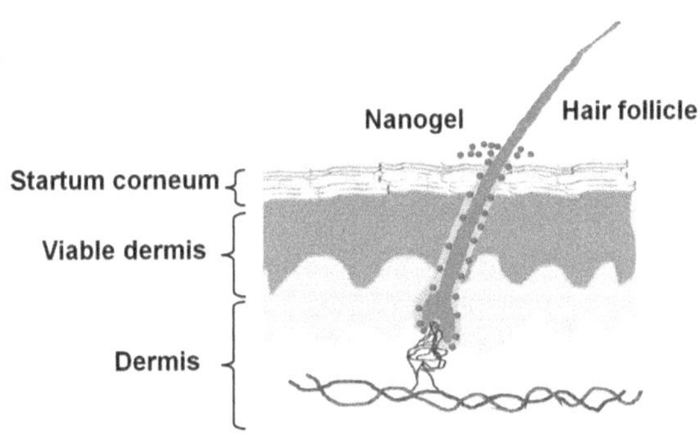

Fig. 9.2 Temperature-responsive nanogel for the follicular delivery of therapeutics.

approach has been limited due to the less penetration of the visible light as well as UV light can damage tissues even at low power [20]. Furthermore, the metal nanoparticle (Au and Ag)-incorporated NG will be stimulated in the presence of NIR light, the induced mild hyperthermia results in the phase transition of the NG's network for the sustained delivery of therapeutics not only in vitro but also in vivo [21]. For example, light-responsive, Au-NP-containing NGs have been prepared for use as photothermal therapeutics, and these photo-responsive NGs can be selectively activated for drug release in a specific disease region by externally applied photoirradiation. The results show that photo-responsive NGs containing metal NPs may be useful for both delivery and release of drug as well as localized heating for thermal therapy. This dual mode therapeutic approach based on hybrid NGs has been shown to enhance therapeutic efficacy [22].

Magnetic field-responsive NGs: In this type of exogenous NGs, magnetic field-responsive hybrid NG respond to externally applied magnetic field for the controlled and sustained delivery of therapeutics into specific sites. The accommodation of magnetic nanoparticle into NGs can be performed by emulsion polymerization or in situ synthesis methods. However, magnetic field-responsive NGs have a remarkable potential for use in drug delivery platform like photo-/light-responsive NG [23]. Initially, magnetic nanoparticle including NGs can be directed to the targeted therapeutic site by direct application of a permanent magnetic field, and once accumulated in the target site, the hybrid NGs release their therapeutic payloads upon irradiation. For example, Bleomycin-loaded magnetic field-responsive hybrid NGs were used to treat squamous cell carcinoma in rabbits [24]. A permanent magnet was placed directly over the tumor surface for 1 day after NG injection. Over the next two weeks, the tumor shrank significantly in size due to magnetic field site-directed NG accumulation and drug release within the tumor. Moreover, magnetic field-responsive hybrid NGs can also be used for photothermal therapy, since it can generate heat (mild hyperthermia) under magnetic field. For example, in the case of PNIPAM including magnetic hybrid NG, the significant NG deswelling occurred, and this promoted the release of encapsulated doxorubicin (DOx) [25].

9.4.2 Indigenous stimuli-responsive NGs

In this section, we have described the common indigenous stimuli like pH, redox, and enzyme-responsive NGs for the controlled delivery of therapeutics.

pH-responsive NGs: pH variations in the cancer cells (intracellular and extracellular compartments) have been significantly exploited for the development of advanced polymeric nanocarriers during the past few decades. The composition of polymer which contains either ionizable groups or acid-sensitive groups will undergo conformational changes/selective cleavage in response to environmental pH variation in the intra/extracellular compartments of the cancer cells [11, 12].

The pH-responsive NGs would alter their confirmation/size under respective pH conditions, which reflects the pH microenvironment of the specific cell site whereas the deliverable drug is anticipated to be released. While designing the cationic pH-responsive NGs, the pH should be less than its pK_a, or they can be designed to be swellable anionic NGs ($pH > pK_a$). For example, Cindy et al. reported pH-sensitive hydrogel system consisting of poly(methacrylic acid-*g*-ethylene glycol) (P(MAA-g-EG)) and acryloyl group modified-cholesterol-bearing pullulan NGs were developed for the controlled delivery of an anticonvulsant drug, pregabalin [26]. The swelling and drug release behavior of this pH-responsive NG synthesized by different photopolymerization approaches at visible and UV light wavelengths were studied at acidic and basic pH values. Manchun et al. reported that DOx-loaded detain NG for the controlled delivery, apart from the other works they have shown that therefore to evaluate the accelerated (25 °C/60% RH) and long-term (5°C) stability of DNGs prepared with formaldehyde (FD) and glyoxal (GDNGs) as cross-linker by determining the change in their physicochemical properties. The mean diameter decreased with time during long-term storage. The drug content between freshly prepared (initial day) and after storage at 5°C for 180 days of doxorubicin (DOx-loaded FDNGs and DOx-loaded GDNGs) was not significantly different ($P > 0.05$), but decreased after storage under the accelerated condition [27].

Redox-responsive NGs: Redox-responsive NGs are formulated by using either chemical cross-linking of thiol-containing polymer/monomer or disulfide-bearing polymer/monomer. However, the intracellular glutathione concentration of cancer cells was significantly high as compared to normal cells, which leads to the destabilization of the NG for controlled and sustained delivery of therapeutic payloads [28]. Recently, Park et al. reported NGs smaller than 100 nm by a simple core cross-linking reaction of PSI-*g*-PEG micelles with diamines and subsequent chemical conversion of hydrophobic core to hydrophilic polyaspartate. The exposure of these NGs to the reducing agent results in the cleavage of the disulfide cross-links of the NGs, which then undergo complete structural disintegration and produce completely water-soluble residues. The NGs showed an accelerated release of drugs at the reductive intracellular conditions, and Dox was efficiently translocated to the nucleus of cancer cells (Fig. 9.3) [29]. Evidently, we can tune the properties of the NG by tethering the respective monomer while preparing the NG, Miao et al. have reported that a redox-responsive polyionic liquid-based NGs enhance the redox-responsive properties of the NG. The results show that drug-loaded NGs were further oxidized to form disulfide bonds again, which could prevent leakage of model drug from NGs. Only less than 50% RhB could be released in the absence of DTT, while over 80% RhB could be released with the addition of reducing agent. Apart from that, a plenty of work has been reported related to the controlled delivery of therapeutics by using biopolymer/disulfide and synthetic polymer/disulfide compositions [30].

Enzyme-responsive NGs: Enzyme-responsive nanocarriers have been significantly explored for the biomedical application including drug delivery and sensing; the intracellular enzymes played a pivotal role for the destabilization of the NG, which leads to the controlled and sustained delivery of therapeutic cargo not only in vitro but also in vivo. Recently, Wang et al. have reported that an enzyme-stimuli NG based on G4 PAMAM dendrimers was constructed as a nanocarrier for drug delivery. The NG showed much higher drug-loading capacity and the sustained Dox release properties triggered by the enzyme as compared to free Dox with a burst releasing manner in vitro. The nanocarriers were nontoxic and well biocompatible, while the NG1 loaded with Dox achieved the same cytotoxicity as free Dox. Such a dendrimer-based NG may be a potential nanocarrier for the drug delivery in cancer therapy [31]. In order to improve the quality of the NGs, a dual-enzyme loaded multifunctional hybrid NG has been reported for highly

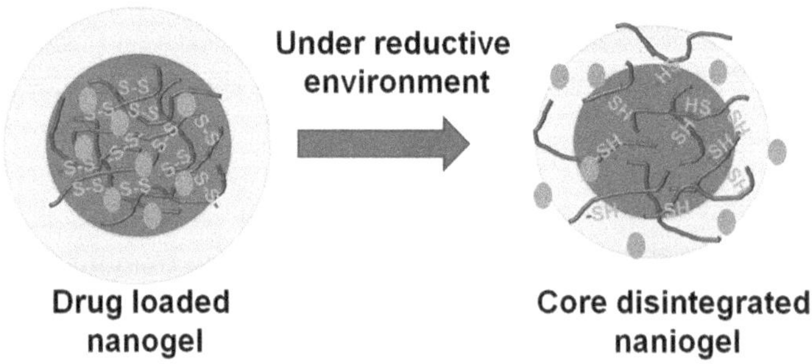

Fig. 9.3 Reduction-responsive nanogel for the delivery of therapeutics.

efficient pathological responsive US (ultrasound) and T2-weighted MR (magnetic resonance) imaging. The favorable design of the SGC agent facilitates a fast and effective enzymatic reaction to generate a sufficient amount of oxygen (O_2) bubbles for contrast in US imaging. This SGC agent also exhibits increased accumulation in the acidic environment of the targeted areas and ensures encapsulation of the SPIO particles within the hydrogel in order to significantly enhance transverse relaxation rates and generate strong MR signals [32].

9.5 Conclusion and future perspective

Interestingly, a plethora of stimuli-responsive NG has been reported for the controlled and sustained delivery of therapeutics in anticancer therapy. Even though several studies have been carried out for the developments of NG very few results can be taken from bench to bedside in anticancer therapy. In this regard, there are few points which have to be focused more as follows: (i) the stability and controlled accommodation of payload without leakage, (ii) the specificity of NGs to the targeted sites, (iii) multicomponent precursor for enhancing the release of payloads, and (iv) preparation of consistent batch with similar dosage.

References

[1] M. Vicario-de-la-Torre, J. Forcada, Gels 3 (2017) 16.
[2] K.S. Soni, S.S. Desale, T.K. Bronich, J. Control. Release 240 (2016) 109.
[3] D. Li, C.F. van Nostrum, E. Mastrobattista, T. Vermonden, W.E. Hennink, J. Control. Release 259 (2017) 16.
[4] G. Soni, K.S. Yadav, Saudi Pharm. J. 24 (2016) 133.
[5] D.M. Eckmann, R.J. Composto, A. Tsourkas, V.R. Muzykantov, J. Mater. Chem. B 2 (2014) 8085.
[6] M.M. Yallapu, M. Jaggi, S. Chauhan, Drug Discov. Today 16 (2011) 457.
[7] F. Sultana, Manirujjaman, M. Imran-Ul-Haque, M. Arafat, S. Sharmin, J. Appl. Pharm. Sci. 3 (2013) S95.
[8] H.K.S. Yadav, N.A.A. Halabi, G.A. Alsalloum, J. Pharm. Pharm. Sci. 1 (2017) 1.
[9] Y. Nomura, Y. Sasaki, M. Takagi, T. Narita, Y. Aoyama, K. Akiyoshi, Biomacromolecules 6 (2005) 447.
[10] H.A. Aliyar, P.D. Hamilton, E.E. Remsen, N. Ravi, J. Bioact. Compat. Polym. 20 (2005) 169.
[11] S. Mura, J. Nicolas, P. Couvreur, *Nat. Mater.* 12 (2013) 991.
[12] J.V. John, R.P. Johnson, M.S. Heo, B.K. Moon, S.J. Byeon, I. Kim, J. Biomed. Nanotechnol. 11 (2015) 1.
[13] J. Ramos, A. Imaz, J. Forcad, Polym. Chem. 3 (2012) 852.
[14] C.-Y. Quan, H. Wei, Y.-X. Sun, S.-X. Cheng, K. Shen, Z.-W. Gu, X.-Z. Zhang, R.-X. Zhuo, J. Nanosci. Nanotechnol. 8 (2008) 2377.

[15] T. Kawano, Y. Niidome, T. Mori, Y. Katayama, T. Niidome, Bioconjug. Chem. 20 (2009) 209.
[16] F.F. Sahle, M. Giulbudagian, J. Bergueiro, J. Lademann, M. Calderón, Nanoscale 9 (2017) 172.
[17] F. Rancan, M. Giulbudagian, J. Jurisch, U. Blume-Peytavi, M. Calderón, A. Vogt, Eur. J. Pharm. Biopharm. 116 (2017) 4.
[18] Y. Gao, S.C.F. Au-Yeng, C. Wu, Macromolecules 32 (1999) 3674.
[19] O. Zavgorodnya, C.A. Carmona-Moran, V. Kozlovskaya, F. Liu, T.M. Wick, E. Kharlampieva, J. Colloid Interface Sci. 506 (2017) 589.
[20] S. Pioge, A. Nesterenko, G. Brotons, S. Pascual, L. Fontaine, C. Gaillard, E. Nicol, Macromolecules 44 (2011) 594.
[21] M. Molina, M. Asadian-Birjand, J. Balach, J. Bergueiro, E. Miceli, M. Calderón, Chem. Soc. Rev. 44 (2015) 6161.
[22] Z. Zhang, J. Wang, X. Nie, T. Wen, Y. Ji, X. Wu, Y. Zhao, C. Chen, J. Am. Chem. Soc. 136 (2014) 7317.
[23] S. Laurent, D. Forge, M. Port, A. Roch, C. Robic, L. Vander Elst, R.N. Muller, Chem. Rev. 108 (2008) 2064.
[24] K. Adriane, J. Huang, G. Ding, J. Chen, Y. Liu, J. Drug Target. 14 (2006) 243.
[25] G. Ciofani, C. Riggio, V. Raffa, A. Menciassi, A, Cuschieri, Med. Hypotheses 73 (2009) 80.
[26] G.E. Cinay, P. Erkoc, M. Alipour, Y. Hashimoto, Y. Sasaki, K. Akiyoshi, S. Kizile, *ACS Biomater.* Sci. Eng. 3 (2017) 370.
[27] S. Manchun, C.R. Dass, P. Sriamornsak, Asian J. Pharm. 11 (2016) 648.
[28] J.F. Quinn, M.R. Whittaker, T.P. Davis, Polym. Chem. 8 (2017) 97.
[29] C.W. Park, H.-M. Yang, M.-A. Woo, K.S. Lee, J.-D. Kim, J. Ind. Eng. Chem. 45 (2017) 182.
[30] K. Rahimian, Y. Wen, J.K. Oh, Polymer 72 (2015) 387.
[31] Y. Wang, Y. Luo, Q. Zhao, Z. Wang, Z. Xu, X. Jia, ACS Appl. Mater. Interfaces 8 (2016) 19899.
[32] X. Wang, D. Niu, P. Li, Q. Wu, X. Bo, B. Liu, S. Bao, T. Su, H. Xu, Q. Wang, ACS Nano 9 (2015) 5646.

CHAPTER 10

Stimuli-responsive nano drug delivery systems for anticancer therapy

Reju George Thomas, Yong Yeon Jeong
Department of Radiology, Chonnam National University Hwasun Hospital, Hwasun, South Korea

Contents

10.1 Introduction

Cancer is the foremost cause of death in humans worldwide. General chemotherapy strategies to overcome cancer have certain limitations such as side effects caused due to unintended accumulation in normal cells rather than cancer cells [1]. The advent of nanomedicine has ushered a new era of targeted therapeutics which are highly effective for curing cancer. The featured mechanism of nanoparticle for targeting tumor region is enhanced by permeation and retention (EPR) effect [2]. Size, shape, and surface chemistry of nanoparticle determine its ability to accumulate in tumor by EPR effect [3]. There are also other factors that govern the nanoparticle uptake and subsequent release of drug to tumor site.

Biomimetic Nanoengineered Materials for Advanced Drug Delivery
https://doi.org/10.1016/B978-0-12-814944-7.00010-2

Stimuli-responsive nanoparticles as drug delivery systems are a widely used methodology for delivering therapeutic drug molecules to the tumor site without affecting the surrounding area [4]. Cancer chemotherapy by stimuli-responsive nanoparticle has gained popularity in recent times, with the advancement of nanotechnology, material chemistry, and nanomedicine [5, 6] Stimuli-responsive nanoparticles are classified into two categories: (1) internal (pH, redox, and enzyme) and (2) external (radiation, thermal, and electromagnetic) stimuli-responsive nanoparticles based on the mechanism of inducing drug delivery (Fig. 10.1) [7, 8]. Other features that guide stimuli-responsive nanoparticles are the time of action, released amount of drug, and multiple dosage capability [9, 10]. Material, design, and chemistry of nanoparticles also play a key role in spatiotemporal controllable drug delivery apart from the stimuli causing factors [11, 12].

The main advantage of external and internal stimuli-responsive nanoparticle system is the rapid and controlled release of drug by tuning nanoparticle property [13]. However, there are a few limitations for external stimuli-

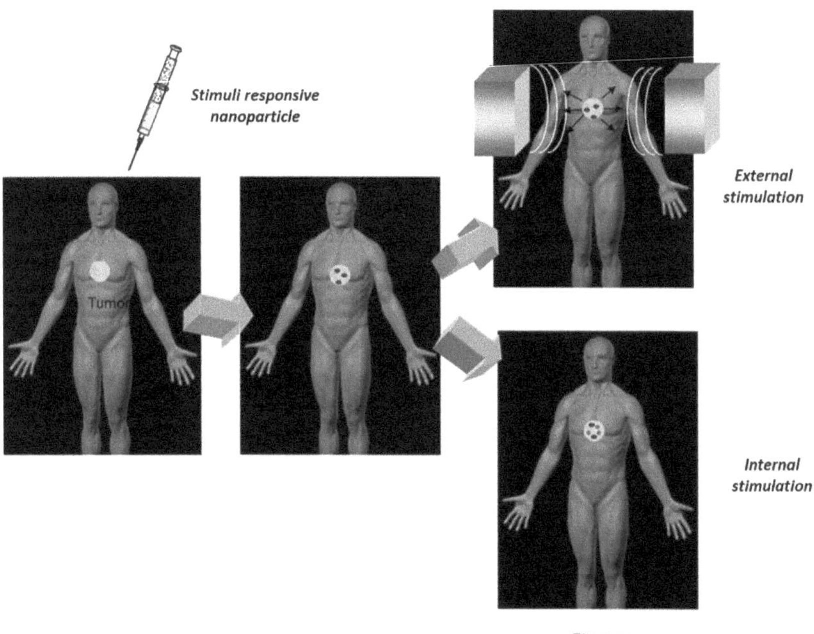

Fig. 10.1 Schematic illustration of stimuli-responsive drug delivery systems. *Modified and reprinted with permission from S.F. Medeiros, A.M. Santos, H. Fessi, A. Elaissari, Stimuli-responsive magnetic particles for biomedical applications, Int. J. Pharm. 403(1-2) (2011) 139–161. Copyright © 2018 Elsevier B.V.*

responsive material which is the inability of external stimuli to reach the tumor site to induce efficient drug release. Internal stimuli are triggered within the tumor region and therefore does not face the above dilemma. The disadvantage of internal stimuli is that they cannot be controlled by an external agency like the oncologist who calculates the time, amount of drug release, and multidose regimen for the patient.

10.2 Internal stimuli-responsive system

10.2.1 pH stimuli-responsive system

pH stimuli method for releasing drug from nanoparticles is considered to be the most effective method compared with other drug release methods. The pH stimuli-release mechanism is based on the fact that tumor microenvironment is slightly acidic and highly acidic in intracellular compartments (pH 5.5) [14]. Tumor environment is acidic due to the accumulation of lactic acid produced by tumor cells under rapid cell division. This phenomenon is known as Warburg effect [15, 16]. When pH-sensitive nanoparticles are exposed to this acidity a stimuli-responsive action initiates resulting in change in material chemistry and finally drug gets released (Fig. 10.2) [17].

Both organic and inorganic materials are used to make nanoparticles for pH stimuli responsiveness [18]. Polymer-based nanoparticles are widely used for making pH-sensitive systems as the chemistry like solubility, conformation, or volume can be easily manipulated. In a recent study, phospholipid end capped polyurethane-based lipopolymer (PL-PU-PL) was synthesized (80 ± 10 nm). Acetal units were responsible for acid degradable nature and doxorubicin (DOx) was loaded to form a pH-sensitive nanomicelle for tumor treatment. Molecular dynamics (MD) studies showed the structure to be compact with polymer chains forming outside and doxorubicin loaded inside. ^{1}H NMR peak at low pH showed aldehyde peak signifying acetal bond breaking to aldehyde and alcohol. In vitro release of doxorubicin was analyzed with pH 4.0 showing 80%, pH 6.0 showing 22%, and pH 7.4 showing only 8% of cumulative release. Cell uptake and FACS analysis also showed similar pattern with pH 4.0 having maximum release of doxorubicin. In vivo biodistribution showed high tumor accumulation by EPR effect and finally tumor inhibition study in SCC7 tumor model in BALB/c nude mice showed lowest tumor volume for DOx-loaded DCPE-PU-DCPE compared to control groups [19].

Doxorubicin-based PEG-b-PCL nanomicelles (100–150 nm) were developed with varying amounts of β-carboxylic amides on the polyester

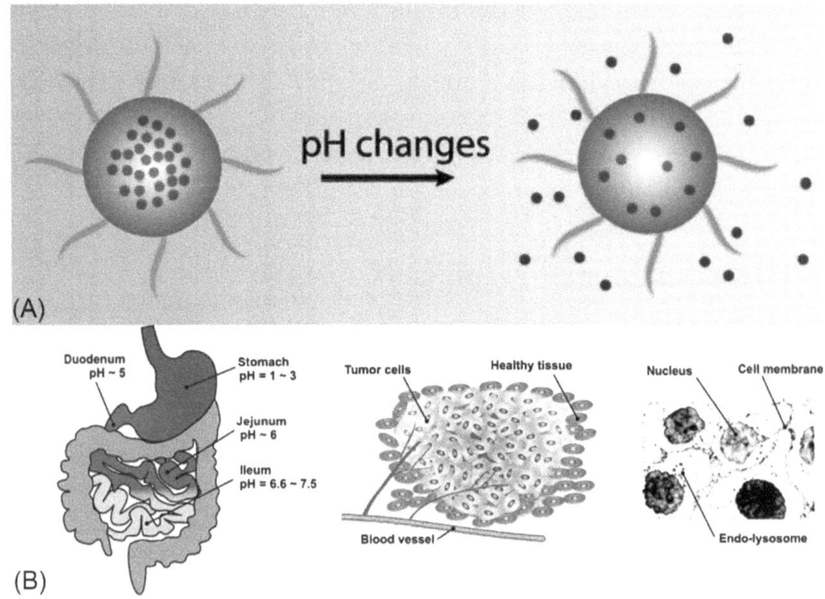

Fig. 10.2 Design of acid-responsive NPs for selective drug release. (A) pH stimuli-responsive release of nanoparticle. (B) Targeting at the organ, tissue, and cellular level: the GI tract is characterized by a pH gradient, solid tumors have a characteristic acidic extracellular environment different from healthy tissues and endo-lysosomes are more acidic in comparison to the cytoplasm (shown in *red—light gray* in print version). *Reprinted with permission from W. Gao, J.M. Chan, O.C. Farokhzad, pH-Responsive nanoparticles for drug delivery, Mol. Pharm. 7(6) (2010) 1913–20. Copyright (2017) American Chemical Society.*

moiety to improve acid labile property [20]. β-carboxylic amides are subjected to hydrolysis at lower pH (pH 6.0) leading to charge reversal of polymeric micelle from negative to positive. An additional advantage of charge reversal is stated to be effective cell uptake in tumor environment which was proved in HepG2 liver cancer cell line. In vitro cytotoxicity studies in HepG2 also proved the same. Similar charge reversal technique was applied in another work using the same β-carboxylic amide property. PLLeu-PLL(DMA)-Tat(SA) (PPDTS) nanomicelles (20 nm) were synthesized with cell nuclei targeting Tat peptide. Tat peptide is amidated to succinyl chloride [Tat (SA)] by click chemistry. The dual pH stimuli strategy for cancer cell killing is described as follows. As the nanomicelles reach tumor environment the pH change from 7.4 to 6.5 will result in charge reversal from negative to positive causing uptake of cancer cells. Once inside cancer cell endocytotic pH of 5.0 will cause hydrolyzation of succinyl chloride, unmasking Tat peptide's positive charge and nuclear targeting function [21].

Doxorubicin-loaded polymeric drug conjugate (polylactide-graft-doxorubicin) was synthesized by click reaction method. The size was 100 nm with 32% DOx loading [22]. Acid-sensitive Schiff-based linkage between DOx and PLA-g-PDA was broken at pH 5.5 and this feature was utilized for stimuli-responsive drug release strategy. In another work a similar polymer drug conjugate of DOx and PEG was made using bis-norbornene as acid-labile linker by brush-first ring-opening metathesis polymerization (ROMP) method [23]. Another interesting strategy for drug release from self-assembled hyaluronic acid (HA) nanoparticle is mineralizing the same using calcium phosphate forming hydroxyapatite nanoparticles loaded with DOx. This system when exposed to low pH can dissolute the mineral thereby releasing drug to tumor microenvironment [24].

HA-conjugated and liposome-supported mesoporous silica nanoparticles (DOx@HA-PL-MSNs) were used as $CD44^{+}$-based targeting system for cancer therapy. In this system pH-sensitive liposome was used to load DOx-containing MSNs which can breakup in acidic environment thereby releasing DOx from MSN cage. This system proved to be highly biocompatible by virtue of bio-friendly lipids, HA, and silica [25]. Ceramic-based nanosystem zirconia was synthesized with dual pH stimuli-mediated activity. Hollow mesoporous zirconia nanospheres showed dual pH-mediated sensitivity by molecular switching function to release DOx payload [26].

10.2.2 Redox stimuli-responsive system

Redox potential of cancer cells is higher than other cells in our body with almost 100- to 1000-fold increase. Glutathione (GSH) is responsible for redox-mediated disulfide linkage destabilization in nanoparticles. The difference between GSH level in blood and tumor environment is almost 500-fold which makes it an effective strategy for internal stimuli-responsive drug release application. Nanoparticle based on redox stimuli-response contains disulfide bond-linked removable shell that can easily shed in the presence of GSH enzyme and release the drug (Fig. 10.3) [27, 28].

PEGylation is a popular strategy to promote longer circulation of nanoparticle for efficient and prolonged drug delivery by EPR effect. But a disadvantage of PEG capping is the resistance for drug to get released in tumor environment. Shedding of PEG shell of nanoparticle is a novel idea applied for controlled release of drug. A PEGylated liposome with orthoester

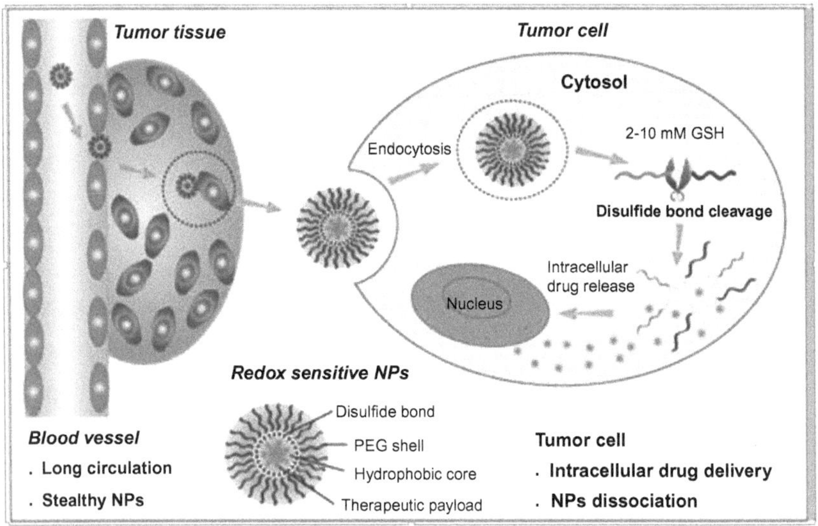

Fig. 10.3 Schematic illustration of redox-sensitive NPs with disulfide-linked PEG shell which can respond to tumor intracellular GSH microenvironments for controlled release of therapeutic agents. *Reprinted with permission from H. Wen, Redox sensitive nanoparticles with disulfide bond linked sheddable shell for intracellular drug delivery, Med. Chem. 4(11) (2014) 748–755. Copyright © 2017 OMICS International.*

linkage was synthesized with redox-sensitive property. In DOx-loaded mPEG-SS-PzLL nanomicelles, poly-e-benzyloxycarbonyl-L-lysine is the lipid part responsible for micelle formation. At increased GSH level, it showed destabilization with the release of DOx. MCF-7 cell lines showed toxicity compared to control [29]. This shell-detachable strategy used in disulfide-linked poly(3-caprolactone)-b-poly(*N*,*N*-dimethylamino-2-ethyl methacrylate) (PCL-SS-PDMA) showed GSH-dependent release of DOx and gene. This positively charged nanoparticles showed cellular uptake capability and DNA-binding property [30].

MSNs with redox response for drug delivery was synthesized which contain gold nanoparticles (AuNPs) as gate keeper (MSN-Fc@Au-CD) [31]. Instead of conventional GSH-reactive disulfide linkage here a system called host-guest interaction where cyclodextrin-modified AuNP bound to MSN control the release of drug was used. Ferrocene (Fc) molecules act as the host which bind to cyclodextrin (guest) and respond to redox reaction by releasing hydrophilic Fc+ thereby opening the gate for drug release from cavities. In another redox-sensitive peptide-functionalized MSN system with Arg-Gly-Asp (RGD) peptide conjugated for capping (DOx@MSN-S-S-RGD), RGD acts as both capping agent and targeting moiety. Therefore the system showed enhanced cellular uptake in COS-7 cells and enhanced the release of DOx [32]. Another peptide-based capping and release strategy in MSN is shown in Ref. [33]. Protein-based capping and release strategy were applied where transferrin-capped MSNs (DOx@MSNs-S-S-Tf) showed cancer-targeting effect and release of DOx by uncapping of transferrin due to disulfide link destabilization. This system boosts slower drug release in the absence of GSH and rapid release in intracellular condition [34].

10.2.3 Enzyme stimuli-responsive systems

Diseased condition like cancer can cause altered expression of enzyme leading to increased concentration of matrix metalloproteinases (MMPs), hyaluronidase, or cathepsin (Fig. 10.4) [35–38]. MMP-2 and MMP-9 enzymes are considered to play an important role in some tumor progression. The usage of MMP in stimuli-responsive behavior of nanoparticles for drug delivery comes when the nanosystem is functionalized with an MMP cleavable component.

For treating hepatocellular carcinoma (HCC), a novel galactosylated (GAL) liposome (Gal-PEG-PD) was synthesized with MMP-2 cleavable component, PEG-peptide with dioleoylphosphatidylethanolamine (PEG-PD) [39]. Gal-PEG-PD liposomes would not be taken up by normal liver cells but only HCC which has high expression of MMP-2 enzyme. N^4-octadecyl-1-β-D-arabinofuranosylcytosine (NOAC) was embedded into the liposomes which have been reported to have anticancer effect. Effectiveness of liposome was tested by cytotoxicity test and it is shown to have high toxicity in HepG2 cell line. In another work, MMP-9-sensitive liposome was made with an MMP-9 activatable component [40]. This is a collagen mimetic peptide that shows uncorking effect in the presence of MMP-9 leading to the release of drug to tumor environment. The effect of the liposomal release of drug was demonstrated using a model drug (carboxyfluorescein) which has fluorescence to indicate the percentage of release.

Polymer-based nanoparticle for stimuli-responsive drug release was demonstrated using PEG-co-PCL nanoparticles [41]. Here high expression of MMP-2 and MMP-9 in glioblastoma and ability of protamine to transport across cancer cellular membrane was utilized for efficient PTX delivery.

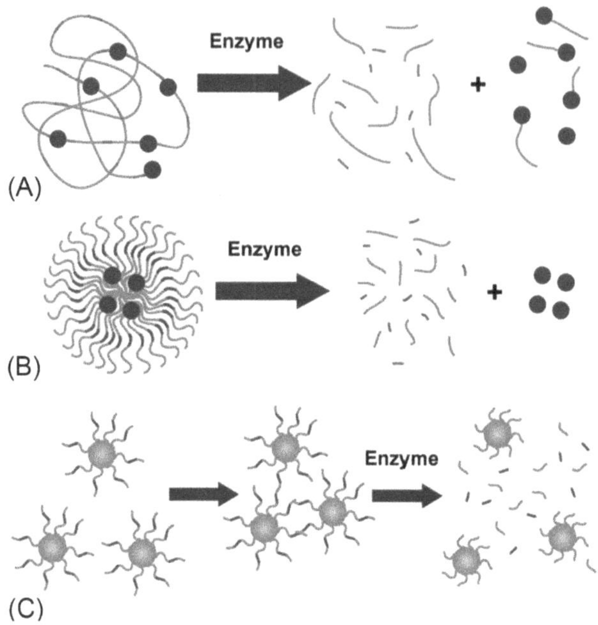

Fig. 10.4 Enzyme-responsive nanomaterials for drug delivery and diagnostics. (A) Polymer-based nanoparticles can be covalently modified with drugs through an enzyme-cleavable linker so that the enzyme activity triggers drug delivery in the tissue of interest. (B) Polymer-stabilized liposomes can be loaded with drugs, whose degradation can be programmed to be triggered by an enzyme. (C) Inorganic nanoparticles can be used for diagnostics when the activity of the target hydrolase controls the assembly or disassembly of the nanoparticles, which in turn changes the physical properties of the nanoparticle solution. *Reprinted with permission from R. de la Rica, D. Aili, M.M. Stevens, Enzyme-responsive nanoparticles for drug release and diagnostics, Adv. Drug Deliv. Rev. 64(11) (2012) 967–78. Copyright © 2012 Elsevier B.V.*

Modified protamine with MMP activatable component (ALMWP) was functionalized by maleimide-thiol coupling chemistry and loaded with PTX. At the vicinity of MMP, the ALMWP of ALMWP-NP is cut and polycationic part LMWP is released and the subsequent uptake of LMWP-NP with PTX occurs in glioblastoma. In vitro test and tumor spheroid tests were conducted to test the uptake and drug release property of the nanoparticle, which showed good cytotoxicity in C6 glioma cells.

Apart from liposomal and polymer systems, MSNs have been considered for making MMP-sensitive nanoparticle for stimuli-responsive drug release. Phenylboronic acid conjugated human serum albumin-grafted MSN was synthesized. MMP-2 substrate peptide and cell-penetrating peptide polyarginine were also included in this nanoparticle [42, 43]. In another work

MSN with hyaluronidase-responsive drug release property was developed. HA was grafted onto MSN by streptavidin-biotin chemistry and DOx was loaded to nanoparticles to demonstrate enzyme-sensitive drug release. HA also functioned as targeting moiety to CD44 overexpressed cell lines. In vitro analysis was done in colon cancer cell lines which showed CD44-specific uptake and toxicity. In vivo study in xenograft model of colon-26 cell line corroborated with in vitro test showing tumor reduction in test group [44].

10.2.4 ROS stimuli-responsive systems

The basis of reactive oxygen species (ROS)-based stimuli-responsive system is on the fact that cancer cells produce more ROS than normal cells [45]. There are several theories stating reasons for high ROS in cancer cells. One of them states that mitochondrial ROS is an important factor for maintaining growth responses in cancer cells [46, 47]. This is the basis for developing ROS stimuli-responsive nanoparticle for cancer therapy (Fig. 10.5).

Recently, PEGylated bilirubin-based nanoparticles (BRNPs) have been developed for exploiting oxidative stress in cancer cells [48]. Bilirubin is an antioxidant that have intrinsic ROS quenching effect. Here on-demand DOx release was possible by the conversion of hydrophobic bilirubin to hydrophilic biliverdin that destabilizes the nanostructure thereby releasing DOx. This effect was shown by exposing BRNPs to peroxy radical generator called 2,2′-azobis(2-amidinopropane) dihydrochloride (AAPH) at 100 mM concentration. The nanoparticles showed disruption proved by size decrease measured by DLS. In vivo anticancer efficacy was tested in lung adenocarcinoma cell (A549) tumors and showed 55% tumor reduction compared to control. Finally, DOx@BRNPs have been proven to be excellent ROS stimuli-responsive nanoparticles.

ROS-responsive nanoparticle was synthesized containing β-cyclodextrin (β-CD) which has low immunogenicity and biocompatibility [49]. 4-Phenylboronic acid pinacol ester (PBAP) was conjugated to β-CD to form a core-shell nanoparticle in the presence of lecithin and poly(ethylene glycol-2000)-grafted distearoylphosphatidylethanolamine (DSPE-PEG) by self-assembly and loaded docetaxel (DTX) to form DTX/Ox-bCD nanoparticle. In vivo anticancer efficacy was tested in melanoma-bearing nude mice established by subcutaneous injection of B16F10 cells, it showed a significant tumor reduction compared to control. In another work, b-cyclodextrin (b-CD) was used to gate MSN encapsulating DOx and

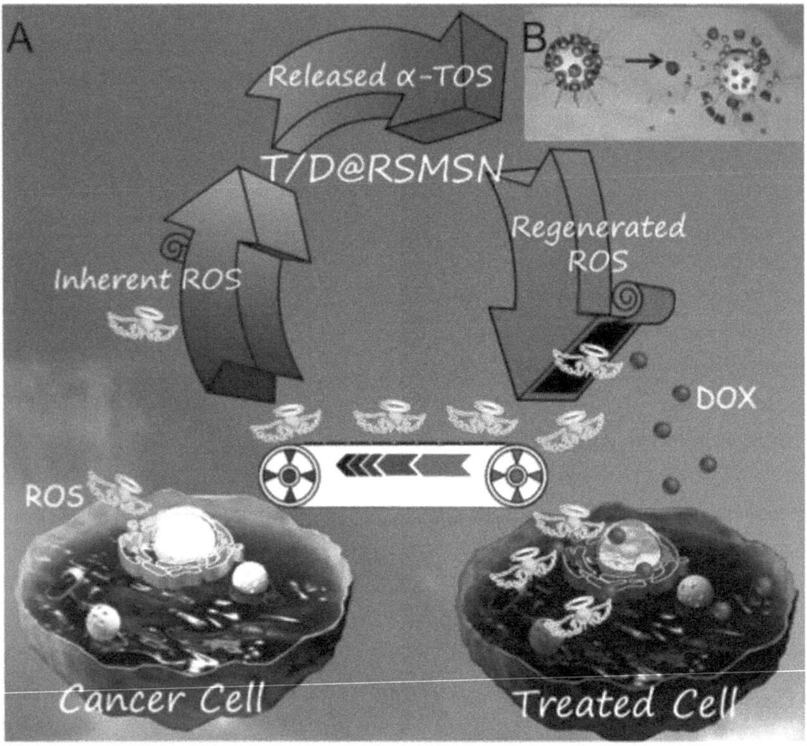

Fig. 10.5 Schema showing ROS stimuli-responsive release of DOx. (A) Inherent ROS acting to release DOx and α-TOS and initiate a positive feedback loop to create more ROS and stimuli response. (B) Schematic structure of T/D@RSMSN. *Reprinted with permission from J.J. Hu, Q. Lei, M.Y. Peng, D.W. Zheng, Y.X. Chen, X.Z. Zhang, A positive feedback strategy for enhanced chemotherapy based on ROS-triggered self-accelerating drug release nanosystem, Biomaterials 128 (2017) 136–146. Copyright © 2017 Elsevier B.V.*

ROS generator α-tocopheryl succinate (α-TOS) by ROS degradable thioketal (TK) linker [50]. ROS generating property of α-TOS was tested by DCF-DA assay where high fluorescence expression of MCF-7 cells uptaking α-TOS/DOx-loaded MSN proved the same. The DOx release by ROS-triggered mechanism was validated using confocal laser scanning microscopy (CLSM) where MCF-7 cells with α-TOS/DOx-loaded MSN showed more DOx fluorescence compared to DOx-loaded MSN. Both in vitro and in vivo experiments were done to prove that α-TOS/DOx-loaded MSN is better than DOx-loaded MSN for treating human breast cancer. In vivo tumor reduction study proved this hypothesis with

Table 10.1 Recent application of internal stimuli responsive nanoparticles.

Component	Payload	Stimulus	Reference
PL-PU-PL	Doxorubicin	pH	[19]
PEG-b-PCL	Doxorubicin	pH	[20]
PLLeu-PLL(DMA)-Tat(SA)	Doxorubicin	pH	[21]
PLA-g-PDA	Doxorubicin	pH	[22]
DOx-M-PEG-HANPs	Doxorubicin	pH	[24]
DOx@HA-PL-MSNs	Doxorubicin	pH	[25]
HMZNs	Doxorubicin	pH	[26]
PCL-SS-PDMA	Doxorubicin	Redox	[30]
MSN-Fc@Au-CD		Redox	[31]
DOx@MSN-S-S-RGD	Doxorubicin	Redox	[32]
DOx@MSNs-S-S-Tf	Doxorubicin	Redox	[34]
Gal-PEG-PD		Enzyme	[39]
PEG-co-PCL	Paclitaxel	Enzyme	[41]
MSN-HA/Dox	Doxorubicin	Enzyme	[44]
DOx@BRNP	Doxorubicin	ROS	[48]
DTX/Ox-bCD	Docetaxel	ROS	[49]
α-TOS/DOx loaded MSN	Doxorubicin	ROS	[50]

significant tumor reduction for α-TOS/DOx-loaded MSN compared to DOx-loaded MSN (Table 10.1).

10.3 External stimuli-responsive system

10.3.1 Photothermal stimuli-responsive system

Photothermal stimuli-responsive chemotherapy is a rising area in the field of combinatorial therapies (Fig. 10.6). Basically a nanoparticle system containing photothermal agent [e.g., gold, ICG (indocyanine green), or heptamethine dye] when irradiated with 808-nm laser initiate a thermal response in nanoparticle system leading to disruption and eventual drug release. Apart from drug release mechanism these system also exhibit synergistic photothermal-mediated hyperthermia to kill cancer cells [51].

Thermosensitive liposomes are popular choice among nanoparticle systems for thermal stimuli drug release (IR780-BTSL-FA) [52]. A bubble-generating liposome (BTSL) was made with thermosensitive function to target folic acid (FA) overexpressing cancer cells. IR-780 was chosen as photothermal agent of choice and DOx/ammonium bicarbonate (NH_4HCO_3) as the drug- and bubble-generating precursor. The mechanism of release was explained by the conversion of NH_4HCO_3 to CO_2

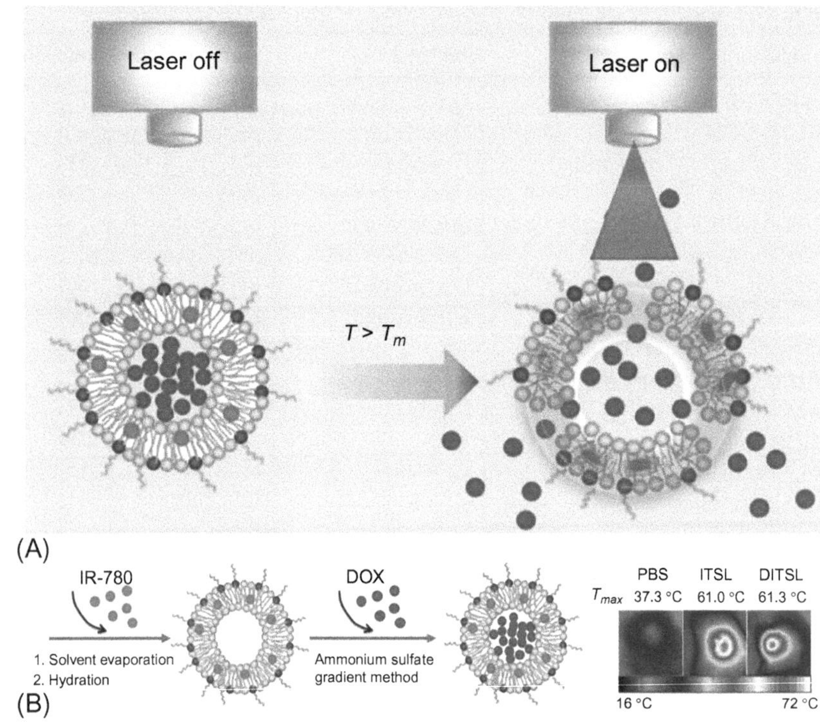

Fig. 10.6 DITSL nanoparticles for PTT/chemotherapy. (A) Schematic diagram of DOx release from DITSL under NIR-laser irradiation. (B) Schematic diagram of DITSL preparation. (C) The representative infrared photothermal images of tumors following laser irradiation. *Reprinted with permission from F. Yan, W. Duan, Y. Li, H. Wu, Y. Zhou, M. Pan, H. Liu, X. Liu, H. Zheng, NIR-laser-controlled drug release from DOX/IR-780-loaded temperature-sensitive-liposomes for chemo-photothermal synergistic tumor therapy, Theranostics 6(13) (2016) 2337–2351. Copyright © 2017 Ivyspring International Publisher.*

bubbles which disrupt the lipid bilayer. When IR-780 gets irradiated with 808-nm laser, the thermal energy released cause the formation of CO_2. This phenomenon was proved in cancer cells where DOx release is observed under irradiation. An in vivo study was conducted which showed significant tumor reduction by combined photothermal and chemotherapeutic actions.

In a separate study, liposome with thermosensitive function was made for dual photothermal therapy and stimuli-responsive release of drug [53]. DOx and IR-780 was loaded in hydrophilic and hydrophobic compartments of the liposome. Laser irradiation study showed release of the drug by virtue of temperature-sensitive lipids used in this system. A polymer-based system for photothermal stimuli response system was made with p(*N*-isopropylacrylamide-co-poly[ethylene glycol] methyl ether acrylate)-block-poly (epsilon-caprolactone) (p(NIPAAM-co-PEGMEA)-b-PCL) [54]. A heat-shock protein inhibitor (17-allylamino-17-demethoxygeldanamycin) was loaded in hydrophobic compartment with a near-infrared dye for photothermal-mediated drug release effect. Thermosensitive component of this nanomicelle system of 200 nm was NIPAAM. RAFT (reversible addition-fragmentation) polymerization technique was used to synthesize the final nanomicelle with NIPAAM, PCL, and PEGMEA components. Lower critical solution temperature (LCST) is the temperature at which the nanosystem undergoes a conformational change releasing the drug. For this nanomicelle LCST was determined to be 43°C. Drug release profile study showed almost 65% release. In vitro cytotoxicity test was done in HCT-116 and found to have close to 50% cell death. In vivo test done in colon cancer model also showed synergistic effect of photothermal and photothermal-mediated drug release chemotherapy with a significant reduction in tumor size.

NIPAAM-based thermosensitive nanoparticle was made with "yolk-shell" configuration for photothermal-mediated targeted drug release and multimodal imaging [55]. In this system, Fe_3O_4 was used as photothermal agent and thermosensitive copolymer poly(N-isopropylacrylamide)-co-1-vinyl-2-pyrrolidone p(NIPAM-co-NVP) as shell. DOx was loaded in between the shell and core causing a stimuli-mediated release when the shell compresses from thermal stress. DOx-Fe_3O_4-PNIPAM had LCST of 41°C. Cumulative release of DOx was shown to be 50% at neutral pH and NIR irradiation. In vivo tumor inhibition rate for DOx-Fe_3O_4-PNIPAM with photothermal stimulation was shown to be 91.5%.

Another NIPAAM-based triblock copolymer system [poly(acrylic acid)-b-poly(*N*-isopropylacrylamide)-b-poly(e-caprolactone)]-SH (PAA-b-PNIPAAm-b-PCL-SH) was made with DOx as the drug payload. Here gold nanorod formed the core of micelle which also acted as the photothermal agent. The polymer system was coated onto gold nanorod by simple dialysis method. The LCST of GNR@polymer-DOx was 42°C and achieve a drug release of 50% at thermally excited state [56]. Single-walled carbon nanotube (SWCNT) is an excellent photothermal agent. DOx was coated onto SWCNT by π–π interaction and NIPAAM polymer was incorporated into have a stimuli-responsive behavior. Further Chitosan-oleic acid (CS-OA) was coated to finally have a hydrodynamic diameter of 100 nm at hyperthermic state. This nanosystem was tested for anticancer effect in bladder cancer model in mice and was found to have excellent tumor reduction property with almost 90% reduction [57].

10.3.2 Photodynamic stimuli-responsive system

Photodynamic stimuli-responsive nanoparticle systems deliver anticancer drug on the basis of its ability to accept the ROS from photosensitizer and destabilize the system to promote drug release [58]. ROS responsive linkers like TK linkers are popular choice for photodynamic stimuli-responsive systems (Fig. 10.7).

Photosensitizer chlorin e6 (Ce6), carboxyl-mPEG, and anticancer drug camptothecin with TK linker were combined to form an upconversion nanoparticle (UCNP)-based nanoparticle (Ce6-CPT-UCNPs) with 30 nm size [59]. UCNP was PEGylated for biocompatibility. After that drug and Ce6 were coated onto UCNP. Fluorescence resonance energy transfer

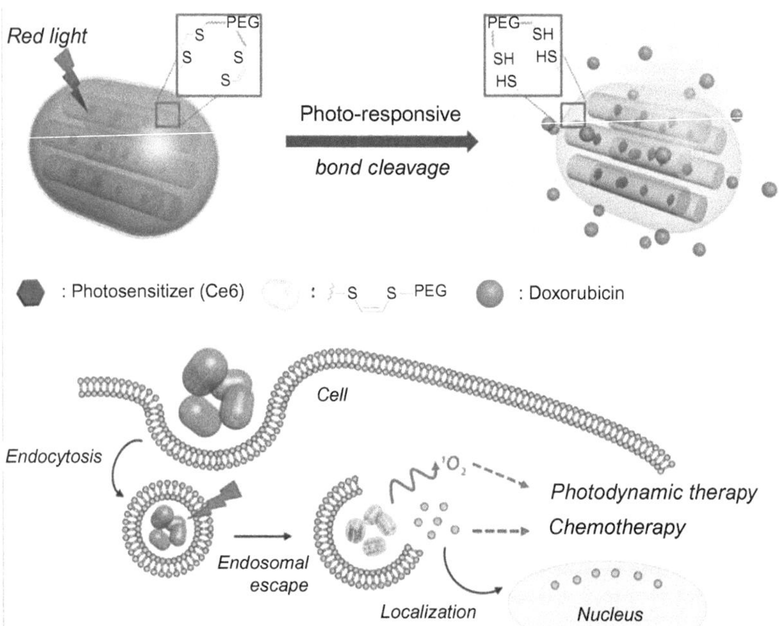

Fig. 10.7 Scheme showing mechanism of photodynamic-mediated DOx release from Ce6-incorporated MSNs and photodynamic-mediated release of drug. *Reprinted with permission from J. Lee, Y.M. Lee, J. Kim, W.J. Kim, Doxorubicin/Ce6-loaded nanoparticle coated with polymer via singlet oxygen-sensitive linker for photodynamically assisted chemotherapy, Nanotheranostics 1(2) (2017) 196–207. Copyright © 2017 Ivyspring International Publisher.*

(FRET) between UCNP and Ce6 is an important feature of this nanoparticle. UCNP emit emission wave of 675 nm which excite Ce6 to emit ROS that breaks TK linker to release the drug. The nanoparticle was tested in vitro in lung cancer cell line (NCI-H460) for cytotoxicity which showed almost 80% cell toxicity under laser irradiation compared to the control. In vivo tumor inhibition study showed almost 100% reduction compared to control.

MSN incorporating Ce6 and DOx with ROS-sensitive PEG covering was synthesized (DOx@CeAP-L-PEG). Singlet oxygen-sensitive labile bis(alkylthio)alkene linker (CeAP-L-PEG) was used as ROS labile covering to protect DOx from unspecific release [60]. After irradiation with laser at 675 nm the PEG covering came off due to photo-responsive bond cleavage of linker and subsequent release of drug happened. Photoresponsive drug release was confirmed by fluorescence assay called singlet oxygen sensor green (SOSG) and intracellular evaluation in HeLa cells after laser irradiation. In vitro cytotoxicity was tested in four cancer cell lines (HeLa, PC-3, Hep3B, and HCT-8) for which cytotoxicity at highest concentration observed was ~60%, ~60%, ~55%, and ~55%, respectively. Finally, in vivo tumor reduction was tested in CT-26 tumor-bearing BALB/c mice; it showed almost complete tumor reduction compared to control mice.

10.3.3 Magnetic stimuli-responsive system

The basis for magnetic stimuli-responsive drug release lies on the ability of the system to utilize its heat-generation property by alternate magnetic frequency (AMF). Since AMF can penetrate through deep tumor tissue this method is excellent for clinical application. In addition to that cancer drug release triggered by AMF hyperthermia will act as synergistic effect to reduce tumor [61–63]. The hyperthermia is generated by AMF in superparamagnetic iron oxide nanoparticles (SPIONs) mainly due to Néel and Brown processes. Briefly, Néel relaxation is caused due to magnetic moment flipping and Brownian relaxation is due to magnetic particle and moment rotation. For bigger ferromagnetic nanoparticles hysteresis loss is also a factor for heat generation. Hysteresis loss is a process where the magnetic domain move and cause friction generating heat [64, 65]. Drug incorporated magnetic nanoparticle rely on the heat-induced destabilization of nanoparticle.

Liposomal nanoparticles (130 nm) were loaded with 10-nm citric acid-coated nanoparticles and DOx (CAMNP@liposomes DOx). Biocompatibility of liposome was tested in L-929 cell lines showing no cytotoxicity.

Hyperthermia and drug release toxicity study were done in CT-26 cell line showing almost 56% cytotoxicity compared to hyperthermia and drug alone [66]. In another study ammonium carbonate containing liposome (MagABC) was made with iron oxide nanoparticle (IONP) and DOx loading. MagABC showed 48 release of DOx at hyperthermia temperature. In cell uptake analysis MagABC showed temperature dependent release of DOx as shown in CLSM images. Cytotoxicity was evaluated in HeLa cells and showed less than 20% viability compared to control groups [67].

Apart from IONP loading, functionalized IONPs are used for stimuli-responsive drug release. In this study IONP (MF66) was functionalized by a cancer-targeting moiety (N6L) and DOx (MF66-N6LDOx) [68]. Nucant multivalent pseudopeptide (N6L) show specific uptake in MDA-MB-231 human breast cancer cell line. MF66 functionalized with N6L and DOx showed uptake and release in MDA-MB-231 as shown by Prussian blue and fluorescence imaging, respectively. The nanoparticle also reports a high specific absorption rate (SAR) of 900 W/g Fe in bare state but 50% reduction in SAR value after functionalization. Maximum DOx release at hyperthermic temperature was reported to be 78% in PBS. In vivo hyperthermia treatment showed significant decrease in tumor size for MF66 with DOx release.

AMF heat generation was utilized for the release of DOx, utilizing the low glass transition temperature of 44°C of polymer which is ideal for drug release [69]. For making this polymeric nanoparticle, the monomer used was pyrrole-3-carboxylic acid (PyCOOH). With the help of a catalyst (iron (III) ion), PyCOOH was polymerized in the presence of DOx and polyvinyl alcohol (PVA) to form DOx/PPyCOOH nanoparticles. IONP formation happened inside the nanoparticle when iron (III) ions converted to iron (II) ions with hydrazine by reduction process. Finally nanoparticle was modified with PEG to increase the bioavailability of nanoparticle and conjugated with FA for active targeting property (Fe_3O_4/DOx/PPy-PEG-FA). The SAR value of Fe_3O_4/DOx/PPy-PEG-FA nanoparticle was evaluated to be 486 W/g which was claimed to be higher than Resovist. The DOx release after AMF application was found to be 90% at 46°C. RPMI8226 multiple myeloma tumor-induced mice were injected with Fe_3O_4/DOx/PPy-PEG-FA nanoparticle intratumorally and AMF was applied for 20 min. At 7 min time point of AMF application temperature of the tumor reached 44°C. There was significant tumor reduction in nanoparticle-treated group due the action of hyperthermia and released DOx (Fig. 10.8).

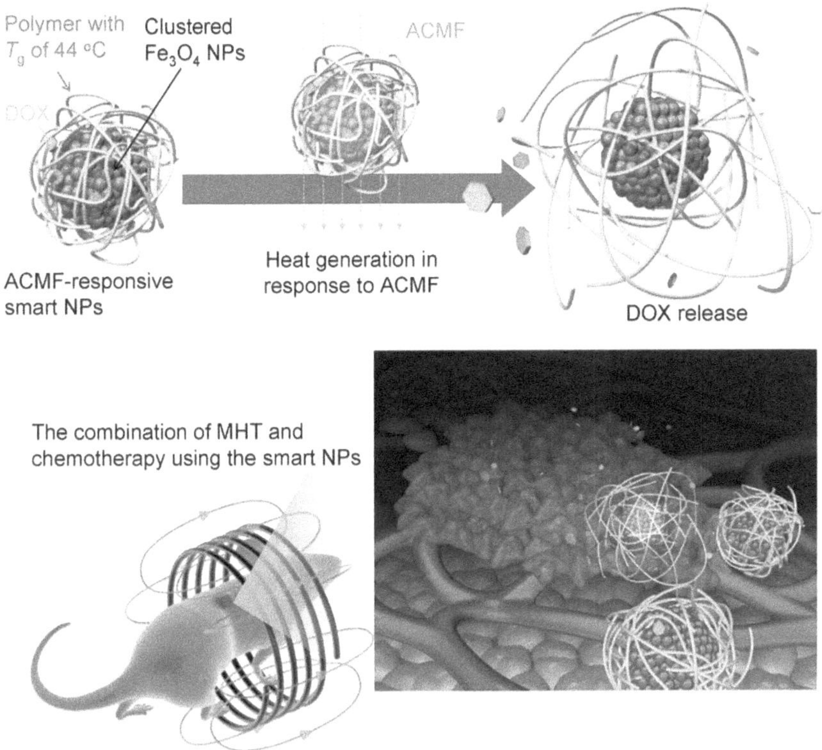

Fig. 10.8 Schema showing magnetic stimuli-responsive mediated drug release. (A) Effect of clustered iron oxide nanoparticles after AMF. (B) Hyperthermia/chemotherapy action for tumor treatment. *Reprinted with permission from K. Hayashi, M. Nakamura, H. Miki, S. Ozaki, M. Abe, T. Matsumoto, W. Sakamoto, T. Yogo, K. Ishimura, Magnetically responsive smart nanoparticles for cancer treatment with a combination of magnetic hyperthermia and remote-control drug release, Theranostics 4(8) (2014) 834–44. Copyright © 2017 Ivyspring International Publisher. Copyright © 2017 Elsevier B.V.*

10.3.4 Ultrasound stimuli-responsive system

Ultrasound stimuli-responsive nano drug delivery system utilizes pressure waves of 20 Hz or higher to induce acoustic cavitation phenomenon in nanomaterial leading to destruction [70]. Another phenomenon called sonoporation also plays an important role in drug uptake inside cancer cells. Sonoporation is a method by which ultrasound frequencies create pores in blood vessels walls leading to extravascular delivery of drug into tumor environment [71, 72].

A polymeric nanoparticle was synthesized that responds to ultrasound and pH to release the anticancer drug cargo to tumor. This nanoparticle is made up of poly(ethylene oxide), (2-tetrahydrofuranyloxy)ethyl methacrylate, and 2-(diethylamino)ethyl methacrylate to form PEO-b-P (DEA-stat-TMA) block copolymer synthesized by atom-transfer radical-polymerization (ATRP) method. This nanoparticle responds to 40 Hz and 180 W ultrasound radiation and low pH to release almost 100% DOx (Fig. 10.9) (Table 10.2) [73].

10.4 Summary and future direction

The primary features of stimuli-responsive drug delivery systems over small-molecule anticancer drugs are their stimuli-responsive drug release, high drug-loading capacity, and biocompatibility. Recent discoveries in the field of nanotechnology have paved way for unique design options for polymeric nanosystems with ability to tune its surface chemistry and physical properties like size and shape to make it responsive to stimuli.

Internal stimuli-responsive system in general relies on pathological difference in microenvironment of cancer and non-cancer area such as pH, GSH level, ROS, and enzymatic variations. The advantage of internal stimuli response is the existence of triggering factor in the diseased area for nanoparticle to release drug. Internal stimuli are triggered within the tumor. However, internal stimuli cannot be controlled by an external agency like the oncologist who calculates the time, amount of drug release, and multidose regimen for the patient.

External stimuli-responsive system depends on external stimuli to trigger the drug release in tumor region. Therefore, advantages are that it is not affected by microenvironmental factors and it is controllable. One limitation of the external stimuli-responsive material is its ability to reach the tumor site to induce efficient drug release. For example, photothermal or photodynamic stimuli have issue of penetration through the body to reach the site of action.

In a long-term perspective, new combinations of polymer, lipid, or inorganic materials have to be synthesized with rational design for effective multi-stimuli-responsive anticancer drug delivery.

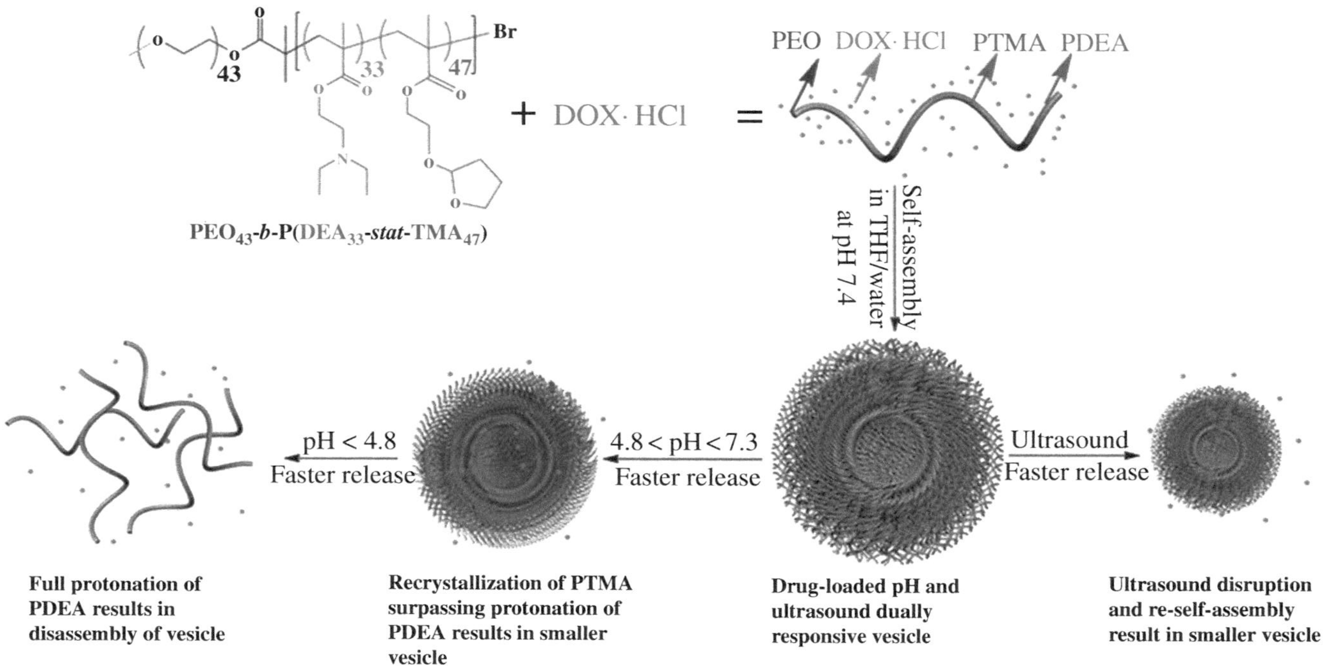

Fig. 10.9 Schema showing ultrasound stimuli-responsive mediated drug release for tumor treatment. *Reprinted with permission from W. Chen, J. Du, Ultrasound and pH dually responsive polymer vesicles for anticancer drug delivery, Sci. Rep. 3 (2013) 2162. Copyright © 2017 Ivyspring International Publisher. Copyright © 2018 Macmillan Publishers Limited.*

Table 10.2 Recent application of external stimuli responsive nanoparticles.

Nanoparticle	Payload	Stimulus	Reference
IR780-BTSL-FA	Doxorubicin	Photothermal	[52]
IR-780/17-AAG micelles	17-AAG	Photothermal	[54]
DOx-Fe3O4-PNIPAM	Doxorubicin	Photothermal	[55]
GNR@polymer-DOx	Doxorubicin	Photothermal	[56]
Ce6-CPT-UCNPs	Camptothecin	Photodynamic	[59]
DOx@CeAP-L-PEG	Doxorubicin	Photodynamic	[60]
CAMNP@liposomes DOx	Doxorubicin	Magnetic	[66]
MagABC	Doxorubicin	Magnetic	[67]
MF66-N6LDOx	Doxorubicin	Magnetic	[68]
DOx/PPyCOOH	Doxorubicin	Magnetic	[69]
Fe_3O_4/DOx/PPy-PEG-FA	Doxorubicin	Magnetic	[70]
PEO-b-P(DEA-stat-TMA)	Doxorubicin	Ultrasound	[73]

Acknowledgment

This research was supported by the Bio & Medical Technology Development Program of the National Research Foundation (NRF) & funded by the Korean government (MSIT) (No. 2019M3E5D1A02068082) Corresponding author: Prof. Yong Yeong Jeong (yjeong@jnu.ac.kr).

References

[1] W. Cheng, L. Gu, W. Ren, Y. Liu, Stimuli-responsive polymers for anti-cancer drug delivery, Mater. Sci. Eng. C Mater. Biol. Appl. 45 (2014) 600–608.
[2] J. Shi, P.W. Kantoff, R. Wooster, O.C. Farokhzad, Cancer nanomedicine: progress, challenges and opportunities, Nat. Rev. Cancer 17 (1) (2017) 20–37.
[3] E. Blanco, H. Shen, M. Ferrari, Principles of nanoparticle design for overcoming biological barriers to drug delivery, Nat. Biotechnol. 33 (9) (2015) 941–951.
[4] H.B. Ruttala, T. Ramasamy, T. Madeshwaran, T.T. Hiep, U. Kandasamy, K.T. Oh, H.G. Choi, C.S. Yong, J.O. Kim, Emerging potential of stimulus-responsive nanosized anticancer drug delivery systems for systemic applications, Arch. Pharm. Res. 41 (2018) 111–129.
[5] V.P. Torchilin, Multifunctional, stimuli-sensitive nanoparticulate systems for drug delivery, Nat. Rev. Drug Discov. 13 (11) (2014) 813–827.
[6] S. Mura, J. Nicolas, P. Couvreur, Stimuli-responsive nanocarriers for drug delivery, Nat. Mater. 12 (11) (2013) 991–1003.
[7] B. Taghizadeh, S. Taranejoo, S.A. Monemian, Z. Salehi Moghaddam, K. Daliri, H. Derakhshankhah, Z. Derakhshani, Classification of stimuli-responsive polymers as anticancer drug delivery systems, Drug Deliv. 22 (2) (2015) 145–155.
[8] J. Yao, J. Feng, J. Chen, External-stimuli responsive systems for cancer theranostic, Asian J. Pharm. Sci. 11 (5) (2016) 585–595.
[9] S. Wang, P. Huang, X. Chen, Stimuli-responsive programmed specific targeting in nanomedicine, ACS Nano 10 (3) (2016) 2991–2994.

[10] J. Du, L.A. Lane, S. Nie, Stimuli-responsive nanoparticles for targeting the tumor microenvironment, J. Control. Release 219 (2015) 205–214.
[11] C. Ding, L. Tong, J. Feng, J. Fu, Recent advances in stimuli-responsive release function drug delivery systems for tumor treatment, Molecules 21 (12) (2016).
[12] C.I. Crucho, Stimuli-responsive polymeric nanoparticles for nanomedicine, ChemMedChem 10 (1) (2015) 24–38.
[13] J. Liao, W. Li, J. Peng, Q. Yang, H. Li, Y. Wei, X. Zhang, Z. Qian, Combined cancer photothermal-chemotherapy based on doxorubicin/gold nanorod-loaded polymersomes, Theranostics 5 (4) (2015) 345–356.
[14] J. Yu, X. Chu, Y. Hou, Stimuli-responsive cancer therapy based on nanoparticles, Chem. Commun. (Camb.) 50 (79) (2014) 11614–11630.
[15] M.G. Vander Heiden, L.C. Cantley, C.B. Thompson, Understanding the Warburg effect: the metabolic requirements of cell proliferation, Science 324 (5930) (2009) 1029–1033.
[16] M.V. Liberti, J.W. Locasale, The Warburg effect: how does it benefit cancer cells? Trends Biochem. Sci. 41 (3) (2016) 211–218.
[17] W. Gao, J.M. Chan, O.C. Farokhzad, pH-Responsive nanoparticles for drug delivery, Mol. Pharm. 7 (6) (2010) 1913–1920.
[18] M. Gisbert-Garzaran, M. Manzano, M. Vallet-Regi, pH-Responsive mesoporous silica and carbon nanoparticles for drug delivery, Bioengineering (Basel) 4 (1) (2017).
[19] J.V. John, R.G. Thomas, H.R. Lee, H. Chen, Y.Y. Jeong, I. Kim, Phospholipid end-capped acid-degradable polyurethane micelles for intracellular delivery of cancer therapeutics, Adv. Healthc. Mater. 5 (15) (2016) 1874–1883.
[20] H. Deng, J. Liu, X. Zhao, Y. Zhang, J. Liu, S. Xu, L. Deng, A. Dong, J. Zhang, PEG-b-PCL copolymer micelles with the ability of pH-controlled negative-to-positive charge reversal for intracellular delivery of doxorubicin, Biomacromolecules 15 (11) (2014) 4281–4292.
[21] S.S. Han, Z.Y. Li, J.Y. Zhu, K. Han, Z.Y. Zeng, W. Hong, W.X. Li, H.Z. Jia, Y. Liu, R.X. Zhuo, X.Z. Zhang, Dual-pH sensitive charge-reversal polypeptide micelles for tumor-triggered targeting uptake and nuclear drug delivery, Small 11 (21) (2015) 2543–2554.
[22] Y. Yu, C.K. Chen, W.C. Law, E. Weinheimer, S. Sengupta, P.N. Prasad, C. Cheng, Polylactide-graft-doxorubicin nanoparticles with precisely controlled drug loading for pH-triggered drug delivery, Biomacromolecules 15 (2) (2014) 524–532.
[23] A.X. Gao, L. Liao, J.A. Johnson, Synthesis of acid-labile PEG and PEG-doxorubicin-conjugate nanoparticles via brush-first ROMP, ACS Macro Lett. 3 (9) (2014) 854–857.
[24] H.S. Han, J. Lee, H.R. Kim, S.Y. Chae, M. Kim, G. Saravanakumar, H.Y. Yoon, D. G. You, H. Ko, K. Kim, I.C. Kwon, J.C. Park, J.H. Park, Robust PEGylated hyaluronic acid nanoparticles as the carrier of doxorubicin: mineralization and its effect on tumor targetability in vivo, J. Control. Release 168 (2) (2013) 105–114.
[25] Z. Wang, Y. Tian, H. Zhang, Y. Qin, D. Li, L. Gan, F. Wu, Using hyaluronic acid-functionalized pH stimuli-responsive mesoporous silica nanoparticles for targeted delivery to CD44-overexpressing cancer cells, Int. J. Nanomedicine 11 (2016) 6485–6497.
[26] M. Wang, G. Gong, J. Feng, T. Wang, C. Ding, B. Zhou, W. Jiang, J. Fu, Dual pH-mediated mechanized hollow zirconia nanospheres, ACS Appl. Mater. Interfaces 8 (35) (2016) 23289–23301.
[27] A.M. Jhaveri, V.P. Torchilin, Multifunctional polymeric micelles for delivery of drugs and siRNA, Front. Pharmacol. 5 (2014) 77.
[28] S. Bauhuber, C. Hozsa, M. Breunig, A. Göpferich, Delivery of nucleic acids via disulfide-based carrier systems, Adv. Mater. 21 (32-33) (2009) 3286–3306.

[29] H.Y. Wen, H.Q. Dong, W.J. Xie, Y.Y. Li, K. Wang, G.M. Pauletti, D.L. Shi, Rapidly disassembling nanomicelles with disulfide-linked PEG shells for glutathione-mediated intracellular drug delivery, Chem. Commun. (Camb.) 47 (12) (2011) 3550–3552.
[30] Y. Li, X. Lei, H. Dong, T. Ren, Sheddable, degradable, cationic micelles enabling drug and gene delivery, RSC Adv. 4 (16) (2014).
[31] H. Qu, L. Yang, J. Yu, T. Dong, M. Rong, J. Zhang, H. Xing, L. Wang, F. Pan, H. Liu, A redox responsive controlled release system using mesoporous silica nanoparticles capped with Au nanoparticles, RSC Adv. 7 (57) (2017) 35704–35710.
[32] Z.-Y. Li, J.-J. Hu, Q. Xu, S. Chen, H.-Z. Jia, Y.-X. Sun, R.-X. Zhuo, X.-Z. Zhang, A redox-responsive drug delivery system based on RGD containing peptide-capped mesoporous silica nanoparticles, J. Mater. Chem. B 3 (1) (2015) 39–44.
[33] D. Xiao, H.-Z. Jia, N. Ma, R.-X. Zhuo, X.-Z. Zhang, A redox-responsive mesoporous silica nanoparticle capped with amphiphilic peptides by self-assembly for cancer targeting drug delivery, Nanoscale 7 (22) (2015) 10071–10077.
[34] X. Chen, H. Sun, J. Hu, X. Han, H. Liu, Y. Hu, Transferrin gated mesoporous silica nanoparticles for redox-responsive and targeted drug delivery, Colloids Surf. B: Biointerfaces 152 (2017) 77–84.
[35] K. Tairong, L. Yarong, G. Tiantian, P. Xiangfang, H. Xianglong, Y. Zhiqiang, Enzyme-responsive nanoparticles for anticancer drug delivery, Curr. Nanosci. 12 (1) (2016) 38–46.
[36] Q. Hu, P.S. Katti, Z. Gu, Enzyme-responsive nanomaterials for controlled drug delivery, Nanoscale 6 (21) (2014) 12273–12286.
[37] A.K. Mitra, V. Agrahari, A. Mandal, K. Cholkar, C. Natarajan, S. Shah, M. Joseph, H.M. Trinh, R. Vaishya, X. Yang, Y. Hao, V. Khurana, D. Pal, Novel delivery approaches for cancer therapeutics, J. Control. Release 219 (Suppl. C) (2015) 248–268.
[38] Y. Yang, J. Aw, K. Chen, F. Liu, P. Padmanabhan, Y. Hou, Z. Cheng, B. Xing, Enzyme-responsive multifunctional magnetic nanoparticles for tumor intracellular drug delivery and imaging, Chem. Asian J. 6 (6) (2011) 1381–1389.
[39] T. Terada, M. Iwai, S. Kawakami, F. Yamashita, M. Hashida, Novel PEG-matrix metalloproteinase-2 cleavable peptide-lipid containing galactosylated liposomes for hepatocellular carcinoma-selective targeting, J. Control. Release 111 (3) (2006) 333–342.
[40] N. Sarkar, J. Banerjee, A.J. Hanson, A.I. Elegbede, T. Rosendahl, A.B. Krueger, A.L. Banerjee, S. Tobwala, R. Wang, X. Lu, S. Mallik, D.K. Srivastava, Matrix metalloproteinase-assisted triggered release of liposomal contents, Bioconjug. Chem. 19 (1) (2008) 57–64.
[41] G. Gu, H. Xia, Q. Hu, Z. Liu, M. Jiang, T. Kang, D. Miao, Y. Tu, Z. Pang, Q. Song, L. Yao, H. Chen, X. Gao, J. Chen, PEG-co-PCL nanoparticles modified with MMP-2/9 activatable low molecular weight protamine for enhanced targeted glioblastoma therapy, Biomaterials 34 (1) (2013) 196–208.
[42] J. Liu, B. Zhang, Z. Luo, X. Ding, J. Li, L. Dai, J. Zhou, X. Zhao, J. Ye, K. Cai, Enzyme responsive mesoporous silica nanoparticles for targeted tumor therapy in vitro and in vivo, Nanoscale 7 (8) (2015) 3614–3626.
[43] Y. Zhao, J.L. Vivero-Escoto, I.I. Slowing, B.G. Trewyn, V.S.Y. Lin, Capped mesoporous silica nanoparticles as stimuli-responsive controlled release systems for intracellular drug/gene delivery, Expert Opin. Drug Deliv. 7 (9) (2010) 1013–1029.
[44] M. Zhang, C. Xu, L. Wen, M.K. Han, B. Xiao, J. Zhou, Y. Zhang, Z. Zhang, E. Viennois, D. Merlin, A hyaluronidase-responsive nanoparticle-based drug delivery system for targeting colon cancer cells, Cancer Res. 76 (24) (2016) 7208–7218.
[45] D. Trachootham, J. Alexandre, P. Huang, Targeting cancer cells by ROS-mediated mechanisms: a radical therapeutic approach? Nat. Rev. Drug Discov. 8 (7) (2009) 579–591.

[46] S.S. Sabharwal, P.T. Schumacker, Mitochondrial ROS in cancer: initiators, amplifiers or an Achilles' heel? Nat. Rev. Cancer 14 (11) (2014) 709–721.
[47] L.B. Sullivan, N.S. Chandel, Mitochondrial reactive oxygen species and cancer, Cancer Metab. 2 (1) (2014) 17.
[48] Y. Lee, S. Lee, D.Y. Lee, B. Yu, W. Miao, S. Jon, Multistimuli-responsive bilirubin nanoparticles for anticancer therapy, Angew. Chem. Int. Ed. Eng. 55 (36) (2016) 10676–10680.
[49] D. Zhang, Y. Wei, K. Chen, X. Zhang, X. Xu, Q. Shi, S. Han, X. Chen, H. Gong, X. Li, J. Zhang, Biocompatible reactive oxygen species (ROS)-responsive nanoparticles as superior drug delivery vehicles, Adv. Healthc. Mater. 4 (1) (2015) 69–76.
[50] J.J. Hu, Q. Lei, M.Y. Peng, D.W. Zheng, Y.X. Chen, X.Z. Zhang, A positive feedback strategy for enhanced chemotherapy based on ROS-triggered self-accelerating drug release nanosystem, Biomaterials 128 (2017) 136–146.
[51] Z. Bao, X. Liu, Y. Liu, H. Liu, K. Zhao, Near-infrared light-responsive inorganic nanomaterials for photothermal therapy, Asian J. Pharm. Sci. 11 (3) (2016) 349–364.
[52] F. Guo, M. Yu, J. Wang, F. Tan, N. Li, Smart IR780 theranostic nanocarrier for tumor-specific therapy: hyperthermia-mediated bubble-generating and folate-targeted liposomes, ACS Appl. Mater. Interfaces 7 (37) (2015) 20556–20567.
[53] F. Yan, W. Duan, Y. Li, H. Wu, Y. Zhou, M. Pan, H. Liu, X. Liu, H. Zheng, NIR-laser-controlled drug release from DOX/IR-780-loaded temperature-sensitive-liposomes for chemo-photothermal synergistic tumor therapy, Theranostics 6 (13) (2016) 2337–2351.
[54] C.L. Peng, Y.I. Chen, H.J. Liu, P.C. Lee, T.Y. Luo, M.J. Shieh, A novel temperature-responsive micelle for enhancing combination therapy, Int. J. Nanomedicine 11 (2016) 3357–3369.
[55] S. Shen, B. Ding, S. Zhang, X. Qi, K. Wang, J. Tian, Y. Yan, Y. Ge, L. Wu, Near-infrared light-responsive nanoparticles with thermosensitive yolk-shell structure for multimodal imaging and chemo-photothermal therapy of tumor, Nanomedicine 13 (5) (2017) 1607–1616.
[56] M. Abbasian, F. Mahmoodzadeh, R. Salehi, A. Amirshaghaghi, Chemo-photothermal therapy of cancer cells using gold nanorod-cored stimuli-responsive triblock copolymer, New J. Chem. 41 (21) (2017) 12777–12788.
[57] H. Gao, Y. Bi, J. Chen, L. Peng, K. Wen, P. Ji, W. Ren, X. Li, N. Zhang, J. Gao, Z. Chai, Y. Hu, Near-infrared light-triggered switchable nanoparticles for targeted chemo/photothermal cancer therapy, ACS Appl. Mater. Interfaces 8 (24) (2016) 15103–15112.
[58] Z. Fang, Z. Bin, Z. Ying, W. Yudong, W. Hanjie, C. Jin, Construction of near-infrared light-triggered reactive oxygen species-sensitive (UCN/SiO2-RB+DOX)@PPADT nanoparticles for simultaneous chemotherapy and photodynamic therapy, Nanotechnology 27 (23) (2016)235601.
[59] C. Yue, C. Zhang, G. Alfranca, Y. Yang, X. Jiang, Y. Yang, F. Pan, J.M. de la Fuente, D. Cui, Near-infrared light triggered ROS-activated theranostic platform based on Ce6-CPT-UCNPs for simultaneous fluorescence imaging and chemo-photodynamic combined therapy, Theranostics 6 (4) (2016) 456–469.
[60] J. Lee, Y.M. Lee, J. Kim, W.J. Kim, Doxorubicin/Ce6-loaded nanoparticle coated with polymer via singlet oxygen-sensitive linker for photodynamically assisted chemotherapy, Nanotheranostics 1 (2) (2017) 196–207.
[61] J.H. Lee, J.W. Kim, J. Cheon, Magnetic nanoparticles for multi-imaging and drug delivery, Mol. Cell 35 (4) (2013) 274–284.
[62] M. Wu, S. Huang, Magnetic nanoparticles in cancer diagnosis, drug delivery and treatment, Mol. Clin. Oncol. 7 (5) (2017) 738–746.

[63] V. Indu, P. Sebastian, D. Alexandra, B. Jeromy, E. Herbert, L. Andreas, Magnetic field-enhanced cellular uptake of doxorubicin loaded magnetic nanoparticles for tumor treatment, Mater. Res. Express 3 (9) (2016)095010.
[64] R.R. Shah, T.P. Davis, A.L. Glover, D.E. Nikles, C.S. Brazel, Impact of magnetic field parameters and iron oxide nanoparticle properties on heat generation for use in magnetic hyperthermia, J. Magn. Magn. Mater. 387 (2015) 96–106.
[65] J. Pearce, A. Giustini, R. Stigliano, P. Jack Hoopes, Magnetic heating of nanoparticles: the importance of particle clustering to achieve therapeutic temperatures, J. Nanotechnol. Eng. Med. 4 (1) (2013) 110071–1100714.
[66] A. Hardiansyah, L.-Y. Huang, M.-C. Yang, T.-Y. Liu, S.-C. Tsai, C.-Y. Yang, C.-Y. Kuo, T.-Y. Chan, H.-M. Zou, W.-N. Lian, C.-H. Lin, Magnetic liposomes for colorectal cancer cells therapy by high-frequency magnetic field treatment, Nanoscale Res. Lett. 9 (1) (2014) 497.
[67] M. Dai, C. Wu, H.M. Fang, L. Li, J.B. Yan, D.L. Zeng, T. Zou, Thermo-responsive magnetic liposomes for hyperthermia-triggered local drug delivery, J. Microencapsul. 34 (4) (2017) 408–415.
[68] S. Kossatz, J. Grandke, P. Couleaud, A. Latorre, A. Aires, K. Crosbie-Staunton, R. Ludwig, H. Dahring, V. Ettelt, A. Lazaro-Carrillo, M. Calero, M. Sader, J. Courty, Y. Volkov, A. Prina-Mello, A. Villanueva, A. Somoza, A.L. Cortajarena, R. Miranda, I. Hilger, Efficient treatment of breast cancer xenografts with multifunctionalized iron oxide nanoparticles combining magnetic hyperthermia and anti-cancer drug delivery, Breast Cancer Res. 17 (2015) 66.
[69] K. Hayashi, M. Nakamura, H. Miki, S. Ozaki, M. Abe, T. Matsumoto, W. Sakamoto, T. Yogo, K. Ishimura, Magnetically responsive smart nanoparticles for cancer treatment with a combination of magnetic hyperthermia and remote-control drug release, Theranostics 4 (8) (2014) 834–844.
[70] Y.Z. Zhao, L.N. Du, C.T. Lu, Y.G. Jin, S.P. Ge, Potential and problems in ultrasound-responsive drug delivery systems, Int. J. Nanomedicine 8 (2013) 1621–1633.
[71] S. Mullick Chowdhury, T. Lee, J.K. Willmann, Ultrasound-guided drug delivery in cancer, Ultrasonography 36 (3) (2017) 171–184.
[72] Z. Fan, R.E. Kumon, C.X. Deng, Mechanisms of microbubble-facilitated sonoporation for drug and gene delivery, Ther. Deliv. 5 (4) (2014) 467–486.
[73] W. Chen, J. Du, Ultrasound and pH dually responsive polymer vesicles for anticancer drug delivery, Sci. Rep. 3 (2013) 2162.

Further reading

[74] S.F. Medeiros, A.M. Santos, H. Fessi, A. Elaissari, Stimuli-responsive magnetic particles for biomedical applications, Int. J. Pharm. 403 (1-2) (2011) 139–161.
[75] H. Wen, Redox sensitive nanoparticles with disulfide bond linked sheddable shell for intracellular drug delivery, Med. Chem. 4 (11) (2014) 748–755.
[76] R. de la Rica, D. Aili, M.M. Stevens, Enzyme-responsive nanoparticles for drug release and diagnostics, Adv. Drug Deliv. Rev. 64 (11) (2012) 967–978.

CHAPTER 11

Graphene-based drug delivery systems

Renu Geetha Bai, Ghaleb A. Husseini
Department of Chemical Engineering, American University of Sharjah, Sharjah, United Arab Emirates

Contents

11.1 Introduction

The process of administering a pharmaceutical compound, in a living organism, to obtain a therapeutic effect is known as drug delivery. Conventional medicine has some precincts such as poor bioavailability, weak targeting specificity, systemic and organ toxicity, etc.; however, modern medicine has created a variety of delivery systems for efficient delivery of therapeutics to overcome these limitations. These advanced drug delivery systems are composed of nanomaterials, polymeric carriers, proteins, complex biomacromolecules, liposomes, proliposomes, microspheres, gels, prodrugs, and lipids that regulate the natural functioning in the body, etc. [1, 2]. Nanoparticles are materials that lie in the size range of 1–100 nm. They exist in numerous shapes such as spheres, rods, wires, planes, etc. The physical and chemical properties of nanomaterials are entirely different from that of their bulk counterparts due to the uniqueness of these nano-formations.

High surface-to-volume ratio, high surface energy, exceptional mechanical, thermal, electrical, magnetic, and optical features are some of the characteristics observed in nanomaterials. These outstanding properties enable an extensive range of applications, such as electronics, energy storage, hybrid materials, medicine, etc. The chemical reactivity and dispensability of

Biomimetic Nanoengineered Materials for Advanced Drug Delivery
https://doi.org/10.1016/B978-0-12-814944-7.00011-4

nanomaterials in various solvents can be controlled by surface modifications. Applications of nanotechnology for treatment, diagnosis, monitoring, and control of biological systems are referred to as nanomedicine by the National Institutes of Health. The ultimate aim of nanomedicine is the delivery and targeting of pharmaceutical, therapeutic, and diagnostic agents which enables specific targeting and thus more efficient therapy while minimizing the side effects. Nanomedicine is a multidisciplinary field of science, in which nanotechnology enables the precise designing and formulation of therapeutics with efficient carriers for site-specific drug delivery and diagnosis [3, 4]. Moreover, the origin of many diseases are the abnormalities at the molecular level where most of the cellular and subcellular organelles are in the nano-range, thus the incorporation of nanosized therapeutics have an upper hand in the treatment of diseases where the blood-brain barrier creates a barrier/challenge to deliver the agents to the infected site. Nanomedical devices are designed to ensure the transport of diagnostic and therapeutic entities, through biological barriers, to efficiently reach the specified target location and treat the disease by molecular interactions [5–7]. Nanomedicine consists of the diagnostic, therapeutic, and theranostic fields of nanomaterials. For nanodrug delivery, a carrier material is employed to overcome cellular barriers, nanocarriers enter through mechanisms of cell internalization by different types of phagocytic and non-phagocytic routes [8]. The rapid growth and development of nanomedicine products holds great promise to improve therapeutic schemes against a variety of diseases and their early detection by diagnosing the biomarkers [9, 10].

Cancer is one of the leading causes of death worldwide. Compared with current cancer therapeutics, nanomedicine offers improved treatment modalities and diagnostic tools. Moreover, a wide range of nanomaterials has been successfully investigated at the preclinical and clinical levels. Carbon nanomaterials, especially graphene and its derivatives, are excellent biomaterials used in nanomedicine. In this chapter, we discuss more about graphene-based drug delivery systems that have been investigated in preclinical and clinical studies [11].

11.2 Graphene and derivatives—characteristics, properties, and applications

Graphene is the two-dimensional (2D), honey-comb crystal lattice carbon structure made of a flat monolayer of carbon atoms. Graphene is the basic building block of many other graphitic structures like zero-dimensional

(0D) fullerene, one-dimensional (1D) carbon nanotubes, and three-dimensional (3D) graphite. Hence, graphene is called the mother of all graphitic structures. Characteristic features of graphene such as unique planar structure, outstanding electric/electronic/thermal/optical/magnetic properties and easy biofunctionalization enable its use in different biological applications including drug delivery, diagnosis, tissue engineering, graphene-quenched fluorescence, graphene-enhanced cell differentiation and growth, graphene-assisted laser desorption, etc. [12]. The exclusive physical, chemical, and mechanical properties of graphene are high surface area (2630 $m^2\ g^{-1}$), outstanding Young's modulus (~1.0 TPa), high thermal conductivity (~5000 $W m^{-1}\ K^{-1}$), great optical transmittance (~97.7%), and exceptional electrical conductivity [13–15]. The major graphene derivatives are graphene oxide (GO), reduced graphene oxide (RGO), graphene quantum dots (GQDs), and graphene nanosheets (GNs). Graphene nano-onions (GNOs), graphene nano-ribbons (GNRs), graphene nano-platelets (GNPs), ultrathin graphite, monolayer graphene, few-layer graphene, etc. The most important derivative of graphene is GO which is the oxidized form of graphene. Apart from graphene, GO possess excellent aqueous processability, amphiphilicity, surface functionalization capacity, surface-enhanced Raman scattering (SERS), fluorescence quenching ability, etc., [16]. RGO is produced by the reduction of GO. Graphite is the 3D form of graphene naturally occurring as igneous rocks, it is cheap and available in large quantities. However, graphite exfoliation to yield individual graphene sheets is not an easy process [17]. Based on the number of graphene sheets, the properties of graphene materials vary. FLG is composed of 2–10 graphene layers, whereas ultrathin graphite comprises more than 10 sheets with thickness up to 100 nm. GO is an oxidized form of chemically modified graphite and nano-GO (NGO) is GO with small lateral dimensions, normally less than 100 nm. GNO is a spherical-shaped concentric layer of graphene, while GNR is a ribbon-shaped graphene stacks fabricated from multiwalled carbon nanotubes. GNP is a disc-shaped multiwalled graphene, synthesized from graphite. GQDs are the 0D form of graphene with the sizes in the range of 3–20 nm [18].

Nanomaterials are synthesized by top-down and bottom-up approaches. RGO is synthesized by the reduction of GO via chemical reduction, photocatalytic reduction, electrochemical reduction, solvothermal reduction, sonochemical reduction, phytochemical reduction, multistep reduction, etc. During reduction, GO undergoes multiple physicochemical changes in structure, electrical conductivity, hydrophilicity, color, and reduction in the side functional groups. Additionally, the healing of defects, restoration of long-range conjugated structures, and the chemical/thermal deoxygenation are other changes that occur during the reduction of GO. As a result of the changes, the properties of RGO varies from GO. Being an excellent biocompatible carbon material, graphene-based nanomaterials are utilized in a comprehensive range of bioapplications including delivery of small molecule drugs/genes, biosensing, tissue engineering, bioimaging, photothermal and photodynamic therapies, etc. [19–25].

11.3 Graphene nanomaterials in nanomedicine

The nanomedicine field offers numerous opportunities to diagnose and treat complicated diseases like cancer, in a timely, more effective manner, and with fewer side effects. Nanomaterials are investigated in medical applications to introduce multimodality diagnosis and therapy of highly complicated diseases. Advanced nanomaterials are used as optical, magnetic, thermal, radioactive imaging agents and contrast agents for optical imaging, photoacoustic imaging, magnetic resonance imaging (MRI), computed tomography (CT), positron emission tomography (PET), and single-photon emission computerized tomography (SPECT). Similarly, nanomaterials are used in biosensor development utilizing fluorescent and plasmonic sensors for nanotherapies including but not limited to chemo-drug delivery, gene/protein delivery, photodynamic therapy, photothermal therapy, etc., to enhance therapeutics efficacy. The toxicity concerns of these nanomaterials are addressed to ensure the safe use of biocompatible materials for medical uses. The combined function of therapy and diagnostics by a single nanoformulation is called nanotheranostics, which offers a great platform for future nanomedicine [26].

Graphene and derivatives are extensively used in nanomedicine in the field of advanced medicine. The role of graphene nanomaterials in the current medical scenarios includes drug delivery, gene/protein therapy, tissue engineering, biosensing, bioimaging, photo/sonotherapy, regenerating scaffolds, antibacterial effects, and biomaterial implants. The compatibility of the material is tested and improved via suitable surface modifications before being applied in the biomedical field. Normally, graphene nanomaterials undergo polymeric surface modifications to make it compatible with different diagnostic and therapeutic applications [27, 28]. The modes of interaction between the graphene nanomaterial and biomolecules depend on the graphene nanomaterial properties, including the number of layers, lateral size, stiffness, hydrophobicity, surface functionalization, dose, etc. [29].

11.4 Nanodrug delivery concept

Drug delivery systems include particulate carriers, composed mainly of polymers, lipids, or other materials to carry therapeutics to improve the pharmacological properties of traditional "free" drugs. Drug delivery systems are designed to regulate the action and biodistribution of drugs, and function as drug reservoirs. Free drug administration faced many challenges such as poor solubility, extravasation associated with normal tissue damage, loss of drug activity in physiological conditions, fast clearance mechanisms, poor biodistribution, and lack of selectivity toward target tissues. Nanodrug delivery systems have resolved these issues to an extent. Nanosize-drug delivery systems generally focus on formulating bioactive molecules in biocompatible nanosystems such as nanocrystals, solid lipid nanoparticles, nanostructure lipid carriers, lipid drug conjugates, nanoliposomes, dendrimers, nanoshells, emulsions, nanotubes, quantum dots, etc. Major drug delivery systems such as liposomes, micelles, and polymeric vesicles have both hydrophilic and hydrophobic natures which improves the drug solubility. Drug delivery systems are useful in avoiding the extravasation associated with tissue damage and premature drug degradation by the programmed sustained release profile process. In traditional medicine, the major antitumor therapeutic approaches include surgery, chemotherapy, radiotherapy, and hormone therapy. A nanodrug delivery system has many advantages as illustrated in Fig. 11.1. The nonspecific targeting of cancer cells has made these approaches noneffective requiring higher doses of drugs to reach the tumor region for maximum efficacy. Nanoparticles can deliver drugs to tumor cells with minimum drug uptake by healthy cells. The conjugation of nanoparticles with ligands (overexpressed on the surface of tumor cells) is a potent site-specific therapeutic approach to treat cancer with high efficacy. The conjugation of nanocarriers with molecules such as antibodies, peptides, nucleic aptamers, vitamins, and carbohydrates can lead to effective targeted drug delivery to cancer cells, leading potentially to cancer attenuation [30].

Due to the surface modifications, size, and biocompatibility of drug delivery systems, they remain in the bloodstream for longer periods of time avoiding renal clearance. Moreover, due to the ligand-mediated targeting and small size, these drug delivery systems have improved uptake by tumors due to the enhanced permeability and retention effect [1]. Moreover, parallel advances

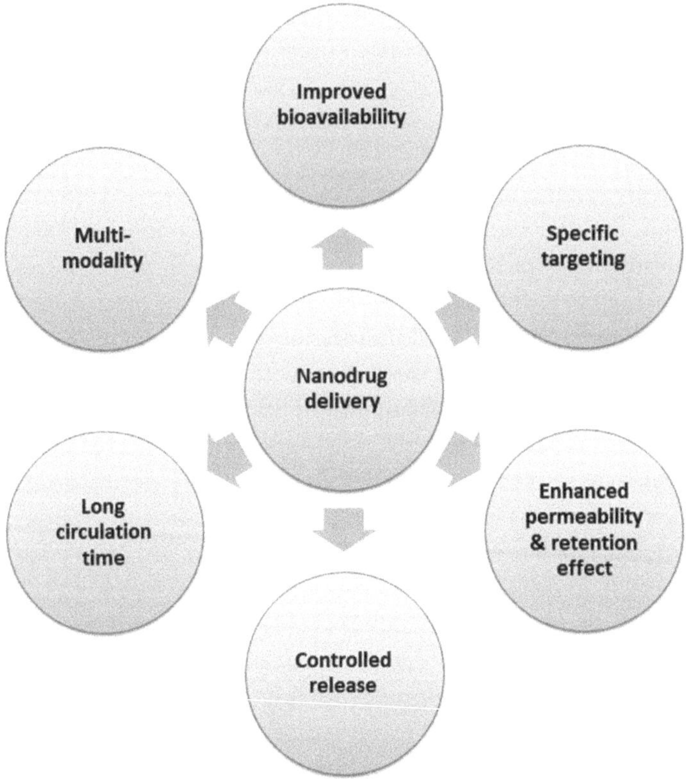

Fig. 11.1 Advantages of nanodrug delivery.

in drug delivery research enabled stimuli-responsive drug release from smart nanodrug carriers [31, 32]. Ultimately, these multifunctional nanomaterials are capable of performing a dual function, namely drug delivery and imaging, that is, tracking their real-time biodistribution and accumulation in target cells/tissues via the nanotheranostic biomaterials [33].

11.5 Graphene materials in drug delivery

The past few decades have witnessed numerous investigations on graphene and its derivatives as nanodrug delivery carriers. The planar structure with extended surface area, high loading capacity, chemical and mechanical stability, superb conductivity, and good biocompatibility of graphene nanomaterials rendered them as efficient carriers for a variety of biomolecules such as

DNA, antibodies, proteins, genes, and small drug molecules [34–37]. Moreover, graphene nanomaterials respond to several stimuli including changes in temperature, pH, magnetic field, electric fields, sound, and radiation. Graphene-based nanomaterial drug delivery systems depicted in Fig. 11.2 are smart systems which enable controlled delivery of drugs through exogenous (temperature, magnetic fields, ultrasound, and light and electric fields) or endogenous stimuli (pH, enzyme concentration, and redox) [38, 39]. These approaches increase drug bioavailability, improve capacity to cross physical barriers, reduce the drug dosage, and limit the toxic side effects of the delivered drugs. The increase in the number of investigations dealing with smart graphene nanomaterials as drug delivery systems proves their efficiency and promise for future clinical use [40].

One-side functionalized GNs known as Janus-type graphenes are used as carriers to deliver camptothecin (CPT), a drug practically insoluble in water; graphene sheets were functionalized with hydrophilic catechol-bearing pyrrolidine rings, while the nonmodified side remains hydrophobic. These partially amphiphilic GNs, dispersed in water, were self-organized in bilayer superstructures, with hydrophilic outer surface and hydrophobic internal space. The later can host hydrophobic molecules such as anticancer drugs and could be used in drug delivery systems in the delivery of CPT [41]. The methotrexate (MTX) anticancer drug was immobilized on a glassy

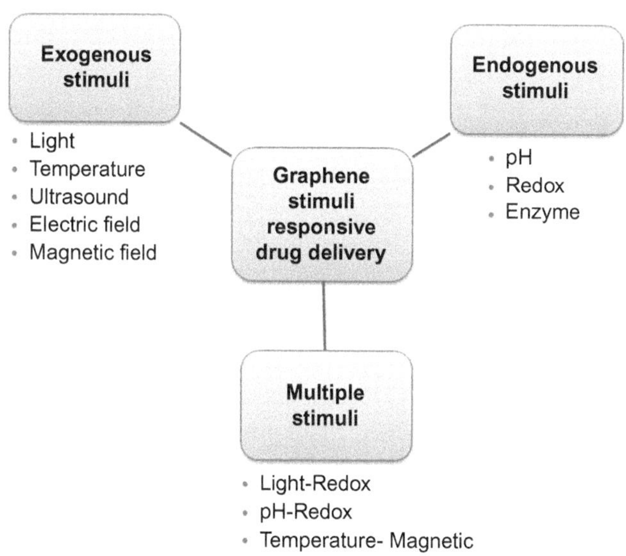

Fig. 11.2 Stimuli-responsive drug delivery in graphene nanomaterials.

carbon-graphene surface to analyze the drug-carrying efficiency of graphene nanomaterials in vitro against 4T1 cancer cells, which showed successful carrier properties and cellular interaction via an electrochemical method [42]. Aliabadi et al. developed a pH-sensitive nanodrug delivery system based on GO for anticancer drug 5-fluorouacil [43]. The drug-loading (%) and drug-entrapment efficiency (%) were calculated to be 29% and 80%, respectively. An in vitro drug release study of nanohybrid graphene in phosphate buffer solution (PBS) showed promising controlled drug release in acidic pH 5.8 (mimicing the tumor environment) up to about 74%, where release was measured to be 59% at pH 7.4, which is a normal physiological pH. This degradation of GO nanohybrid in an acidic medium is thus suitable for pH-based anticancer drug delivery.

Among graphene nanomaterials, GO is the most extensively used derivative due to the presence of side functional groups, hydrophilic nature, and high functionalization capacity. For the efficient delivery of CPT, GO carriers were compared by two different polymer modifications to analyze the suitable carrier. Carrier 1 is surface modified with a natural polymer chitosan and carrier 2 by synthetic polymer polyvinylpyrrolidone. The two GO carriers were also functionalized by folic acid. The comparative study of the biocompatibility of these two nanocarriers were evaluated via hemolysis and anti-inflammatory studies, and the cellular toxicity were tested by 3-(4,5-dimethylthiazol-2-yl)-2,5-diphenyltetrazolium bromide (MTT) and sulforhodamine B (SRB) assays against the MCF-7 cell lines. Chitosan polymerized GO nanocarrier were found to be more compatible toward drug delivery applications with increased drug-loading capacity and a higher MCF-7 inhibition percent [44].

For developing a doxorubicin (DOX) delivery system, GO was functionalized with adipic acid dihydrazide to introduce amine groups, and sodium alginate (SA) was covalently conjugated to GO by the formation of amide bonds. The pH-responsive drug release profile was observed where the measured drug release rate was significantly higher at the acidic tumor pH of 5.0 compared to release under physiological conditions of pH 6.5 and 7.4. The in vitro evaluation of the drug carrier and drug release was analzsed by MTT assay on HeLa and NIH-3T3 cells and experimental results showed that carrier has no obvious toxicity while GO-SA/DOX exhibited notable cytotoxicity. Cell uptake studies of DOX showed that GO-SA could specifically transport the DOX into HeLa cells overexpressing CD44 receptors and displayed enhanced toxicity to HeLa and NIH-3T3 cells [45]. Another pH-responsive DOX delivery system was designed by Rasoulzadeh et al. Here, the GO nanosheets were successfully embedded on the carboxymethyl cellulose (CMC) hydrogel beads to form CMC/GO for the controlled release of DOX. The π-π stacking interaction between DOX and GO resulted in higher loading capacity. The enhanced release of DOX from hydrogel beads at pH 6.8 compared to pH 7.4 indicated the pH dependence of the hydrogel carrier. Moreover, this biocompatible GO-CMC DOX delivery system has no significant toxicity when tested against SW480 colon cancer cells [46]. Similarly, the GO-CMC pH-responsive DOX delivery system was prepared by Rao et al. The in vitro MTT evaluation of this GO-CMC drug delivery system on human cervical cancer HeLa cells and mouse fibroblast NIH-3T3 cells cytotoxicity of the system showed good biocompatibility and no obvious cytotoxicity [47]. Another galactosylated chitosan-modified GO system is designed for DOX delivery with drug-loading capacity reaching 1.08 mg/mg (drug per polymer). The pH-responsive system remained stable under physiological conditions, and DOX was released in a low pH environment. The cell uptake experiments and a cell proliferation analysis demonstrated that the delivery system had a higher cytotoxicity for HepG2 and SMMC-7721 cells. Compared with CS-GO-DOX, the GC-GO-DOX nanoparticles exhibited higher fluorescence intensity in malignant cells. In vivo antitumor experiments on nude mouse validated that the GC-GO-DOX nanoparticles inhibit tumors better than the CS-GO-DOX nanoparticles and thus constitute a better drug carrier system [48]. Another promising DOX delivery system was developed by Ma et al. utilizing NGO with high drug loading and targeted delivery efficiency [49]. DOX-loaded NGO (NGO/DOX), where the agent is encapsulated within a soy phosphatidylcholine (SPC) cell membrane, and the SPC membrane are modified with a PEGylated lipid-FA conjugate for cancer targeting. The resultant nanohybrids (NGO/DOX@SPC-FA) displayed good stability and biocompatibility, high drug-loading capability, efficient cellular uptake, and controlled drug release. Moreover, the FA-modified nanohybrids (NGO/DOX@SPC-FA) could deliver NGO/DOX to cancer cells with improved delivery and enhanced tumor killing efficacy due to the presence of the FA-targeting motifs. The NGO/DOX@SPC-FA nanohybrids were specifically internalized by FA-positive HeLa cancer cells through both macropinocytosis and clathrin-dependent endocytosis and formed lysosomes in the cytosol. The corresponding in vivo biodistribution study showed that the active FA-mediated targeting mechanism improved tumor-targeting ability of the nanohybrid system demonstrating that NGO/DOX@SPC-FA could significantly inhibit tumor growth and prolong the survival time of mice. Our results suggested that NGO/DOX@SPC-FA holds promise as a novel drug delivery system in future cancer therapy [40].

RGO-based nanomaterials are amenable to noncovalent functionalization through hydrophobic interactions. RGO is noncovalently functionalized with maleimide-containing catechol (dopa-MAL). Thiol-maleimide chemistry allows the facile attachment of thiol-containing molecules and thus the attachment of a cancer cell targeting cyclic peptide, c(RGDfC), to the dopa-MAL-modified RGO as a targeted drug delivery system for DOX. When tested, free DOX has been shown to be more effective at killing the human cervical HeLa cancer cells over human breast adenocarcinoma cancer MDA-MB-231cells, where the DOX-loaded rGO/dopa-MAL-c (RGDfC) nanostructure showed more efficacy targeting and killing MDA-MB-231 cells compared to their HeLa counterparts [50]. Gupta et al. developed a superparamagnetic iron oxide-reduced graphene oxide (Fe_3O_4-RGO) nanohybrid drug delivery system by the one-step coprecipitation method to develop a DOX delivery system with magnetic properties. The presence of well-segregated iron oxide nanoparticles reduced the agglomeration of RGO and maintained its layered structure. The in vitro investigations of DOX-loaded nanohybrid (DOX-Fe_3O_4-RGO) on human cervical cancer HeLa cells showed the maximum inhibition of human cervical cancer cell lines during magnetic field-assisted hyperthermia treatment. The synergistic effect of RGO nanohybrid revealed the potential for cancer cell proliferation prevention (up to 90%) when treated at the concentration of 2 mg/mL for 1 million cells at a fixed AC (alternating current) frequency of 265 kHz for 35 min [51].

GQD-based nanohybrid materials are popular in the biomedical fields due to their unique physicochemical properties and outstanding biocompatibility. In addition to drug delivery, GQDs are widely explored in in vivo imaging and in vitro biosensing applications [52]. A highly dispersible and water-soluble GQD is synthesized by acidic oxidation and exfoliation of multiwalled carbon nanotubes to create a GQD-based biocompatible and traceable drug delivery system for the targeted delivery of DOX to cancer cells. The GQDs were covalently linked to the tumor-targeting module biotin (BTN) for the efficient recognition of BTN receptors overexpressed on cancer cells to enable the targeted delivery of DOX. In vitro investigations of the synthesized carrier (GQD and GQD-BTN) on A549 cells reported a very low toxicity which ensured the biocompatibility of the drug carrier. When compared with the free DOX, the GQD-based DOX delivery system GQD-BTN-DOX showed enhanced cellular uptake, acidic-environment-triggered nuclear internalization, and improved toxicity to the cancer cells [53].

Another GQD-based DOX delivery system was designed using arginine-glycine-aspartic acid (RGD)-conjugated GQDs. This RGD-GQD has the advantages of targeted cancer fluorescence imaging as well as tracking and monitoring drug delivery without the need for external dyes. The fluorescence of GQDs enables the real-time monitoring of the cellular uptake of the DOX-GQDs-RGD nanosystem and the pH-dependent release of DOX. GQDs were proven nontoxic to U251 human glioma cells. As shown in Fig. 11.3A, compared with free DOX, DOX-GQDs-RGD conjugates demonstrated significant cytotoxicity to U251 glioma cells. When GQDs are used as drug carriers, the drug efficacy improved without increasing the concentration of DOX. The cellular uptake results of DOX-GQDs-RGD displayed that DOX and some GQDs penetrated into cell nuclei after 16 h of incubation and resulted in improved cytotoxicity as shown in Fig. 11.3B. The confocal laser scanning microscope images of U251 glioma cells treated with DOX-GQDs-RGD by a 405-nm laser are shown in Fig. 11.3C [54].

Another GQD is conjugated with iron oxide and silicon dioxide to form a nanohybrid (GQD)@Fe_3O_4@SiO_2 for the successful delivery of DOX. Thee conjugation produced a nanoprobe with properties similar to targeted drug delivery nanovehicles, including sensing, dual-modal imaging, and therapy. The folic acid functionalization of the nanoprobes increased the cancer targeting property toward HeLa cells. DOX drug molecules were then loaded on the GQD surface via π-π stacking. The in vitro evaluation of this GQD nanoprobe on HeLa cells displayed MRI, fluorescence imaging of living cells, and intracellular drug release process. The cell viability studies of this luminomagnetic nanoprobe showed the low cytotoxicity of the nanocarriers and the enhanced therapeutic efficacy of DOX-loaded nanoprobe for cancer cells as a potential platform for cancer diagnosis and therapy [55].

In another study, a graphene-based nanodrug delivery system is utilized in the furin-mediated sequential delivery of anticancer cytokine and small-molecule drug DOX. This nanodelivery system enables the sequential delivery of the DOX and the cytokine tumor-necrosis-factor-related apoptosis-inducing ligand toward the nucleus and plasma membrane, respectively. The advantages of this combination therapy include improved therapeutic efficacy, and the reduced possibility of drug

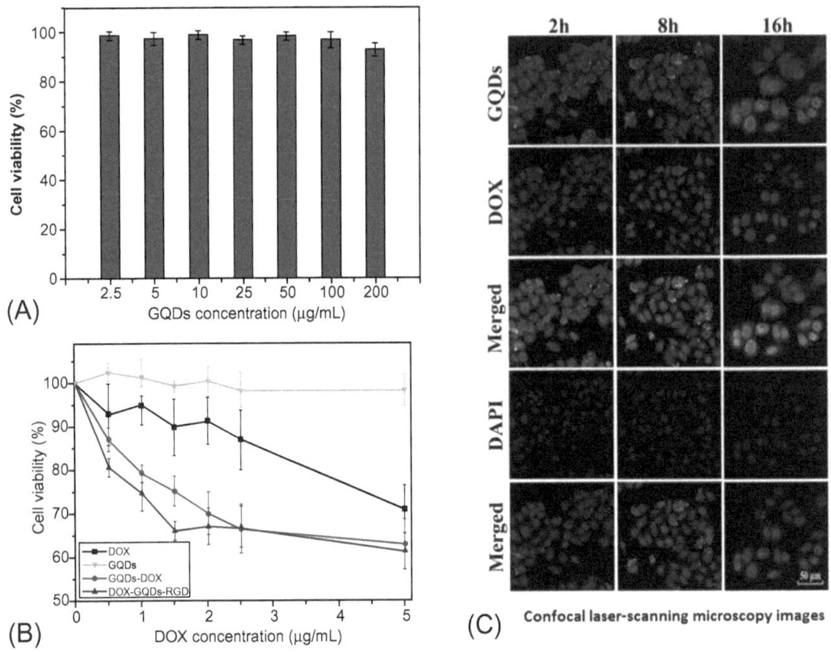

Fig. 11.3 (A) GQDs cell viability analysis of glioma cancer cell line at different concentrations. (B) Cell viability studies of GQDs, free DOX, GQDs-DOX, and DOX-GQDs-RGD at different DOX concentrations. (C) Confocal laser scanning microscope images of U251 glioma cells treated with DOX-GQDs-RGD after incubation for 2, 8, and 16h excited by a 405-nm laser: GQD signals collected in the range of 525±50nm, DOX signals collected in the range of 595±50nm, merged images of GQDs and DOX, cell nuclei staining with DAPI and signals collected from 450±50nm, and merged images of GQDs, DOX, and DAPI. *(Reprinted with permission (license number 4450240357406) from J. Dong, et al., Application of graphene quantum dots for simultaneous fluorescence imaging and tumor-targeted drug delivery. Sensors Actuators B Chem. 256 (2018) 616–623).*

resistance and other side effects. Coencapsulation of this multiple anticancer agents in graphene carrier resulted in a more efficient synergistic antitumor activity [56]. A GO-based codelivery system for the delivery of DOX and small interfering RNA (siRNA) was designed by Sun et al. [57]. The SiRNA could downregulate the expression of the desired gene. Along with the DOX, delivery of SiRNA significantly enhances in vitro and in vivo antitumor effects, which promises a successful application in future clinical trial. Similarly, a natural polymer, chitosan functionalized GO drug delivery system, is utilized for the co-delivery of two anticancer drugs CPT and 3,3′-diindolylmethane (DIM) against breast cancer. The GO system is decorated with folic acid and the drugs vary in their mechanism of action, but they make the promising synergistic effect. Biocompatibility analysis by hemolysis, anti-inflammatory, and cellular toxicity by MTT and SRB assays were performed on the MCF-7 cell lines. In vivo analysis of drug biodistribution, blood biochemical analysis, and drug bioavailability data revealed the enhanced anticancer activity of this synergistic therapy [58]. More examples on graphene drug nanodelivery systems are listed in Table 11.1.

A multifunctional carrier for combined therapy is an efficient way for regulating diseases, where graphene nanomaterials are excellent carriers for multifunctional therapeutic systems. The incorporation of targeting moieties, drug release control biomolecules, fluorescent agent, magnetic nanoparticles, etc., are some of the multifunctional carrier possibilities. A number of multifunctional systems have been developed in graphene nanomaterials for combination therapies, which have proven to be highly efficient in controlling cancer progression. Multifunctional graphene-based magnetic nanohybrid systems were utilized for combined hyperthermia and dual stimuli-responsive drug delivery. The system exhibited high DOX drug loading ($910\,\mu g\,mg^{-1}$) and controlled and repeated drug release where magnetic character supported the hyperthermia and stimuli-responsive drug release for thermo-chemotherapy. This graphene-based magnetic nanocarrier displayed excellent physicochemical, magnetic and biocompatibility properties, and dual exogenous (AC field)/endogenous (pH) stimuli-responsive drug responses for the targeted thermo-chemotherapy, combining magnetic hyperthermia and controlled drug release triggered by the abnormal tumor environment [74]. Similarly, a GO-mesoporous silica system is utilized in the delivery of cisplatin by pH and NIR irradiation-responsive release with fluorescent imaging, as a potential chemo-photothermal agent against cancer cells [75]. Likewise, a multifunctional GQD system is conjugated with folic acid functionalized chitosan to develop a nanocargo for DOX delivery in cancer cells. Folic acid ensured the specific targeting, the polymer functionalization enabled the pH-modulated delivery of DOX, and the inherent fluorescent property of GQDs was utilized in bioimaging. Cellular uptake studies conducted on A549 cell revealed that the GQD nanocargo could specifically transport DOX within cancerous cells [76].

Table 11.1 Graphene nanomaterials in drug delivery.

Graphene Nanomaterial	Drug used	In vitro/in vivo	Remarks	References
GO	DOX	HepG2 cells	Delivery system showed pH-dependent site-specific drug release, excellent encapsulation efficiency, high physiological stability, enhanced cellular uptake, and tunable intracellular drug release profiles	[59]
GO	DOX	HeLa cells	ATP-mediated controlled drug release system, high loading capacity and site-specific drug release	[60]
GO	DOX	LO2, SMMC-7721, mice	High drug stability and external stimuli NIR irradiation responsive drug release	[60]
GO	Dexamethasone	Neurons	Excellent temporal control, dosage flexibility, electrically controlled drug delivery	[39]
GQD	DOX	HeLa cells	Real-time monitoring of the cellular uptake and the consequent release of drugs.	[61]
GQD	DOX	4T1 cells	Multifunctional therapeutic system with synergistic effect, resulting in a higher efficacy to kill cancer cells. Chemo-magnetic hyperthermia therapy or chemo-photothermal therapy	[62]
RGO	Lidocaine hydrochloride	Dermal fibroblast cells, HeLa cells	Subsequent drug release at the desired site by external magnetic stimulation, and potential agent for cancer photothermal therapy. Fluorescent and M.R.I. dual-modality imaging	[63]
GO	DOX	C6 glioma cells	High loading capacity and pH-dependent drug release behavior.	[64]

GO	PTX	Human adrenocortical carcinoma SW-13 cells BALB/c nude mice	Substantial cell killing ability in vitro and tumor suppression ability in vivo with 100% survival rate over 100 days and no relapse	[65]
RGO	PTX	Normal lung MRC-5 cells, nasopharyngeal cancer HK-1cells	Highly stable system, cell-specific cytotoxicity	[66]
RGO	PTX, curcumin	Normal lung, A549 cells, Breast cancer MDA-MB-231 cells	High drug loading capacity, cell-specific cytotoxicity	[67]
RGO	Fluorescein sodium	Hepatic stellate cell lines	High drug-loading efficiency of 82.8%, pH responsive drug delivery	[68]
GO	5-Fluorouracil	Colon cancer cells, mice	Colon-targeted delivery, controlled-release behavior	[69]
GO	Rivastigmine	Human fibroblast cells	Excellent drug entrapment efficiency and controlled release	[70]
GO	DOX	Hela cells	Self-healable, with tuneable adhesion to surfaces, dual stimuli-responsive (temperature and pH) drug delivery	[71]
Nano-GO	5-Iodo-2-deoxyuridine	C6 glioma cells, male wistar rats	Multifunctional nanosystem for diagnostic and therapeutic platforms, a potential way to reduce systemic toxicity	[72]
Nano-GO	Oridonin	esophageal cancer KYSE-30, EC109 cells	Bio-functionalized GO drug delivery nanosystem for enhanced cancer targeting efficiency of drugs	[73]

DOX, Doxorubicin; *PTX*, paclitaxel.

11.6 Challenges

The major challenge faced when using graphene nanomaterials in medical applications is the risk of toxicity. Though graphene nanomaterials are mostly considered biocompatible, there are studies reporting the toxicity effects of graphene materials in microbial, cellular, and in vivo studies. The microbial toxicity of graphene nanomaterials is investigated and GO, RGO, graphite, and graphite oxide were found to have antibacterial activity [77, 78]. Moreover, silver-incorporated graphene nanomaterials also showed excellent species-selective antimicrobial activity towards both Gram-positive and Gram-negative bacteria [10, 79, 80]. The cytotoxicity analysis of graphene nanomaterials depends on size, shape, dose, and duration of exposure. The toxicity analysis in a variety of cell lines displayed mitochondrial injury, plasma membrane damage, DNA fragmentation and chromosomal aberrations, increased intracellular ROS, etc. [81–83]. In vivo evaluations of graphene nanomaterials showed accumulation specifically in the lungs and liver. Mice model studies of short-term exposure of graphene nanomaterial resulted in serious and persistent lung injury, and accumulation in lungs, spleen, and liver which lead to acute injury and inflammation [84–86]. On the contrary, other investigations of graphene nanomaterials on mice, zebra fish, and rat animal models showed limited toxicity and in some cases nontoxicity [87–89].

11.7 Conclusion and future perspectives

Graphene nanomaterials are outstanding carrier materials for drug delivery applications. The specific structure-associated properties enable high loading efficiency, functionalization capacity for targeting, imaging, sensing, stimuli responsive release, etc. Multifunctional drug delivery systems and theranostic systems are the future of graphene-based nanodrug delivery systems. However, more preclinical analysis must be performed to understand the source of toxicity issues associated with graphene nanomaterials. Despite the toxicity concerns and associated biosafety issues, graphene nanomaterials offer an exceptional platform for developing multifunctional nanotherapeutic systems.

References

[1] T.M. Allen, P.R. Cullis, Drug delivery systems: entering the mainstream, Science 303 (5665) (2004) 1818–1822.
[2] G. Tiwari, et al., Drug delivery systems: an updated review, Int. J. Pharm. Investig. 2 (1) (2012) 2.

[3] S.M. Moghimi, A.C. Hunter, J.C. Murray, Nanomedicine: current status and future prospects, FASEB J. 19 (3) (2005) 311–330.
[4] K. Riehemann, et al., Nanomedicine—challenge and perspectives, Angew. Chem. Int. Ed. 48 (5) (2009) 872–897.
[5] B.Y. Kim, J.T. Rutka, W.C. Chan, Nanomedicine, N. Engl. J. Med. 363 (25) (2010) 2434–2443.
[6] R. Duncan, R. Gaspar, Nanomedicine(s) under the microscope, Mol. Pharm. 8 (6) (2011) 2101–2141.
[7] P.N. Prasad, Introduction to Nanomedicine and Nanobioengineering, vol. 1, John Wiley & Sons, 2012.
[8] H. Hillaireau, P. Couvreur, Nanocarriers' entry into the cell: relevance to drug delivery, Cell. Mol. Life Sci. 66 (17) (2009) 2873–2896.
[9] R.G. Bai, et al., Sonochemical and sustainable synthesis of graphene-gold (G-Au) nanocomposites for enzymeless and selective electrochemical detection of nitric oxide, Biosens. Bioelectron. 87 (2017) 622–629.
[10] R.G. Bai, et al., The biogenic synthesis of a reduced graphene oxide–silver (RGO–Ag) nanocomposite and its dual applications as an antibacterial agent and cancer biomarker sensor, RSC Adv. 6 (43) (2016) 36576–36587.
[11] J. Shi, et al., Cancer nanomedicine: progress, challenges and opportunities, Nat. Rev. Cancer 17 (1) (2017) 20.
[12] H. Zhao, et al., Graphene-based nanomaterials for drug and/or gene delivery, bioimaging, and tissue engineering, Drug Discov. Today 22 (9) (2017) 1302–1317.
[13] K. Muthoosamy, R.G. Bai, S. Manickam, Graphene and graphene oxide as a docking station for modern drug delivery system, Curr. Drug Deliv. 11 (6) (2014) 701–718.
[14] A.K. Geim, Graphene: status and prospects, Science 324 (5934) (2009) 1530–1534.
[15] Y. Wang, et al., Graphene and graphene oxide: biofunctionalization and applications in biotechnology, Trends Biotechnol. 29 (5) (2011) 205–212.
[16] A. Gulzar, et al., Bioapplications of graphene constructed functional nanomaterials, Chem. Biol. Interact. 262 (2017) 69–89.
[17] S. Stankovich, et al., Graphene-based composite materials, Nature 442 (7100) (2006) 282.
[18] R.G. Bai, et al., Graphene: a versatile platform for nanotheranostics and tissue engineering, Prog. Mater. Sci. (2017).
[19] K. Ghosal, K. Sarkar, Biomedical applications of graphene nanomaterials and beyond, ACS Biomater. Sci. Eng. 4 (8) (2018) 2653–2703.
[20] H. Bao, et al., Chitosan-functionalized graphene oxide as a nanocarrier for drug and gene delivery, Small 7 (11) (2011) 1569–1578.
[21] L. Wu, et al., Highly sensitive graphene biosensors based on surface plasmon resonance, Opt. Express 18 (14) (2010) 14395–14400.
[22] J.T. Robinson, et al., Ultrasmall reduced graphene oxide with high near-infrared absorbance for photothermal therapy, J. Am. Chem. Soc. 133 (17) (2011) 6825–6831.
[23] J. Ge, et al., A graphene quantum dot photodynamic therapy agent with high singlet oxygen generation, Nat. Commun. 5 (2014) 4596.
[24] S. Sayyar, et al., Covalently linked biocompatible graphene/polycaprolactone composites for tissue engineering, Carbon 52 (2013) 296–304.
[25] S. Zhu, et al., Strongly green-photoluminescent graphene quantum dots for bioimaging applications, Chem. Commun. 47 (24) (2011) 6858–6860.
[26] G. Chen, et al., Nanochemistry and nanomedicine for nanoparticle-based diagnostics and therapy, Chem. Rev. 116 (5) (2016) 2826–2885.
[27] J. Zhou, et al., Tuning electronic and magnetic properties of graphene by surface modification, Appl. Phys. Lett. 95 (10) (2009)103108.
[28] M. Batzill, The surface science of graphene: metal interfaces, CVD synthesis, nanoribbons, chemical modifications, and defects, Surf. Sci. Rep. 67 (3-4) (2012) 83–115.

[29] V.C. Sanchez, et al., Biological interactions of graphene-family nanomaterials: an interdisciplinary review, Chem. Res. Toxicol. 25 (1) (2011) 15–34.
[30] B. Bahrami, et al., Nanoparticles and targeted drug delivery in cancer therapy, Immunol. Lett. 190 (2017) 64–83.
[31] S. Mura, J. Nicolas, P. Couvreur, Stimuli-responsive nanocarriers for drug delivery, Nat. Mater. 12 (11) (2013) 991.
[32] M. Karimi, et al., Smart micro/nanoparticles in stimulus-responsive drug/gene delivery systems, Chem. Soc. Rev. 45 (5) (2016) 1457–1501.
[33] V.P. Torchilin, Multifunctional, stimuli-sensitive nanoparticulate systems for drug delivery, Nat. Rev. Drug Discov. 13 (11) (2014) 813.
[34] Y. Pan, N.G. Sahoo, L. Li, The application of graphene oxide in drug delivery, Expert Opin. Drug Deliv. 9 (11) (2012) 1365–1376.
[35] J. Liu, L. Cui, D. Losic, Graphene and graphene oxide as new nanocarriers for drug delivery applications, Acta Biomater. 9 (12) (2013) 9243–9257.
[36] X. Sun, et al., Nano-graphene oxide for cellular imaging and drug delivery, Nano Res. 1 (3) (2008) 203–212.
[37] Z. Liu, et al., Carbon materials for drug delivery & cancer therapy, Mater. Today 14 (7-8) (2011) 316–323.
[38] K. Yang, L. Feng, Z. Liu, Stimuli responsive drug delivery systems based on nano-graphene for cancer therapy, Adv. Drug Deliv. Rev. 105 (2016) 228–241.
[39] C.L. Weaver, et al., Electrically controlled drug delivery from graphene oxide nano-composite films, ACS Nano 8 (2) (2014) 1834–1843.
[40] X. Yu, et al., Graphene-based smart materials, Nat. Rev. Mater. 2 (9) (2017) 17046.
[41] V. Bekiari, et al., Self-assembly of one-side-functionalized graphene nanosheets in bilayered superstructures for drug delivery, J. Mater. Sci. 53 (16) (2018) 11167–11175.
[42] R.K. Shervedani, et al., Immobilization of methotrexate anticancer drug onto the graphene surface and interaction with calf thymus DNA and 4T1 cancer cells, Bioelectrochemistry 119 (2018) 1–9.
[43] M. Aliabadi, H. Shagholani, A. Gerayeli, Planar polymer-graphene oxide nanohybrid as a 5-fluorouacil carrier in pH-responsive controlled release, J. Drug Deliv. Sci. Technol. 43 (2018) 103–106.
[44] A. Deb, R. Vimala, Natural and synthetic polymer for graphene oxide mediated anticancer drug delivery—a comparative study, Int. J. Biol. Macromol. **107** (2018) 2320–2333.
[45] L. Fan, et al., Sodium alginate conjugated graphene oxide as a new carrier for drug delivery system, Int. J. Biol. Macromol. 93 (2016) 582–590.
[46] M. Rasoulzadeh, H. Namazi, Carboxymethyl cellulose/graphene oxide bio-nanocomposite hydrogel beads as anticancer drug carrier agent, Carbohydr. Polym. 168 (2017) 320–326.
[47] Z. Rao, et al., Carboxymethyl cellulose modified graphene oxide as pH-sensitive drug delivery system, Int. J. Biol. Macromol. 107 (2018) 1184–1192.
[48] C. Wang, et al., Design and evaluation of galactosylated chitosan/graphene oxide nanoparticles as a drug delivery system, J. Colloid Interface Sci. 516 (2018) 332–341.
[49] K. Ma, et al., Cancer cell targeting, controlled drug release and intracellular fate of biomimetic membrane-encapsulated drug-loaded nano-graphene oxide nanohybrids, J. Mater. Chem. B 6 (31) (2018) 5080–5090.
[50] Y. Oz, et al., Functionalization of reduced graphene oxide via thiol–maleimide "click" chemistry: facile fabrication of targeted drug delivery vehicles, ACS Appl. Mater. Interfaces 9 (39) (2017) 34194–34203.
[51] J. Gupta, et al., Superparamagnetic iron oxide-reduced graphene oxide nanohybrid-a vehicle for targeted drug delivery and hyperthermia treatment of cancer, J. Magn. Magn. Mater. 448 (2018) 332–338.

[52] F. Chen, et al., Graphene quantum dots in biomedical applications: recent advances and future challenges, Front. Lab. Med. 1 (2018) 192–199.
[53] D. Iannazzo, et al., Graphene quantum dots for cancer targeted drug delivery, Int. J. Pharm. 518 (1-2) (2017) 185–192.
[54] J. Dong, et al., Application of graphene quantum dots for simultaneous fluorescence imaging and tumor-targeted drug delivery, Sensors Actuators B Chem. 256 (2018) 616–623.
[55] X. Su, et al., A graphene quantum dot@ Fe3O4@ SiO2 based nanoprobe for drug delivery sensing and dual-modal fluorescence and MRI imaging in cancer cells, Biosens. Bioelectron. 92 (2017) 489–495.
[56] T. Jiang, et al., Furin-mediated sequential delivery of anticancer cytokine and small-molecule drug shuttled by graphene, Adv. Mater. 27 (6) (2015) 1021–1028.
[57] Q. Sun, et al., Doxorubicin and anti-VegF sirNa co-delivery via nano-graphene oxide for enhanced cancer therapy in vitro and in vivo, Int. J. Nanomedicine 13 (2018) 3713.
[58] A. Deb, N.G. Andrews, V. Raghavan, Natural polymer functionalized graphene oxide for co-delivery of anticancer drugs: in-vitro and in-vivo, Int. J. Biol. Macromol. 113 (2018) 515–525.
[59] X. Zhao, et al., Design and development of graphene oxide nanoparticle/chitosan hybrids showing pH-sensitive surface charge-reversible ability for efficient intracellular doxorubicin delivery, ACS Appl. Mater. Interfaces 10 (7) (2018) 6608–6617.
[60] R. Mo, et al., ATP-responsive DNA-graphene hybrid nanoaggregates for anticancer drug delivery, Biomaterials 50 (2015) 67–74.
[61] X. Wang, et al., Multifunctional graphene quantum dots for simultaneous targeted cellular imaging and drug delivery, Colloids Surf. B: Biointerfaces 122 (2014) 638–644.
[62] X. Yao, et al., Graphene quantum dots-capped magnetic mesoporous silica nanoparticles as a multifunctional platform for controlled drug delivery, magnetic hyperthermia, and photothermal therapy, Small 13 (2) (2017) 1602225.
[63] R. Justin, et al., Photoluminescent and superparamagnetic reduced graphene oxide–iron oxide quantum dots for dual-modality imaging, drug delivery and photothermal therapy, Carbon 97 (2016) 54–70.
[64] M.-M. Song, et al., Lactoferrin modified graphene oxide iron oxide nanocomposite for glioma-targeted drug delivery, Mater. Sci. Eng. C 77 (2017) 904–911.
[65] W. Deng, et al., Development of biocompatible and VEGF-targeted paclitaxel nano-drugs on albumin and graphene oxide dual-carrier for photothermal-triggered drug delivery in vitro and in vivo, Int. J. Nanomedicine 13 (2018) 439.
[66] R.G. Bai, et al., Acoustic cavitation induced generation of stabilizer-free, extremely stable reduced graphene oxide nanodispersion for efficient delivery of paclitaxel in cancer cells, Ultrason. Sonochem. 36 (2017) 129–138.
[67] K. Muthoosamy, et al., Exceedingly higher co-loading of curcumin and paclitaxel onto polymer-functionalized reduced graphene oxide for highly potent synergistic anticancer treatment, Sci. Rep. 6 (2016) 32808.
[68] K. Chen, et al., Chitosan derivatives/reduced graphene oxide/alginate beads for small-molecule drug delivery, Mater. Sci. Eng. C 69 (2016) 1222–1228.
[69] B. Zhang, et al., A colon targeted drug delivery system based on alginate modificated graphene oxide for colorectal liver metastasis, Mater. Sci. Eng. C 79 (2017) 185–190.
[70] Z. Rahmani, R. Sahraei, M. Ghaemy, Preparation of spherical porous hydrogel beads based on ion-crosslinked gum tragacanth and graphene oxide: study of drug delivery behavior, Carbohydr. Polym. 194 (2018) 34–42.
[71] Y. Chen, et al., Graphene oxide hybrid supramolecular hydrogels with self-healable, bioadhesive and stimuli-responsive properties and drug delivery application, Macromol. Mater. Eng. 303 (2018) 1700660.

[72] S. Shirvalilou, et al., Development of a magnetic nano-graphene oxide carrier for improved glioma-targeted drug delivery and imaging: in vitro and in vivo evaluations, Chem. Biol. Interact. 295 (2018) 97–108.
[73] J.-H. Jiang, et al., Functional graphene oxide as cancer-targeted drug delivery system to selectively induce oesophageal cancer cell apoptosis, Artif. Cells Nanomed. Biotechnol. 46 (2018) 1–11.
[74] R.O. Rodrigues, et al., Multifunctional graphene-based magnetic nanocarriers for combined hyperthermia and dual stimuli-responsive drug delivery, Mater. Sci. Eng. C 93 (2018) 206–217.
[75] A.-V. Tran, et al., Targeted and controlled drug delivery by multifunctional mesoporous silica nanoparticles with internal fluorescent conjugates and external polydopamine and graphene oxide layers, Acta Biomater. 74 (2018) 397–413.
[76] S. De, et al., Tailoring the efficacy of multifunctional biopolymeric graphene oxide quantum dot-based nanomaterial as nanocargo in cancer therapeutic application, ACS Biomater. Sci. Eng. 4 (2) (2018) 514–531.
[77] S. Liu, et al., Antibacterial activity of graphite, graphite oxide, graphene oxide, and reduced graphene oxide: membrane and oxidative stress, ACS Nano 5 (9) (2011) 6971–6980.
[78] S. Liu, et al., Lateral dimension-dependent antibacterial activity of graphene oxide sheets, Langmuir 28 (33) (2012) 12364–12372.
[79] J. Ma, et al., Preparation, characterization and antibacterial properties of silver-modified graphene oxide, J. Mater. Chem. 21 (10) (2011) 3350–3352.
[80] B.-K. Kim, Y.-L. Jo, J.-J. Shim, Preparation and antibacterial activity of silver nanoparticles-decorated graphene composites, J. Supercrit. Fluids 72 (2012) 28–35.
[81] L. Zhang, et al., Functional graphene oxide as a nanocarrier for controlled loading and targeted delivery of mixed anticancer drugs, Small 6 (4) (2010) 537–544.
[82] T. Lammel, et al., Internalization and cytotoxicity of graphene oxide and carboxyl graphene nanoplatelets in the human hepatocellular carcinoma cell line Hep G2, Part. Fibre Toxicol. 10 (1) (2013) 27.
[83] O. Akhavan, E. Ghaderi, A. Akhavan, Size-dependent genotoxicity of graphene nanoplatelets in human stem cells, Biomaterials 33 (32) (2012) 8017–8025.
[84] M.C. Duch, et al., Minimizing oxidation and stable nanoscale dispersion improves the biocompatibility of graphene in the lung, Nano Lett. 11 (12) (2011) 5201–5207.
[85] K. Yang, et al., In vivo pharmacokinetics, long-term biodistribution, and toxicology of PEGylated graphene in mice, ACS Nano 5 (1) (2010) 516–522.
[86] K.P. Wen, et al., Accumulation and toxicity of intravenously-injected functionalized graphene oxide in mice, J. Appl. Toxicol. 35 (10) (2015) 1211–1218.
[87] V.E. Fako, D.Y. Furgeson, Zebrafish as a correlative and predictive model for assessing biomaterial nanotoxicity, Adv. Drug Deliv. Rev. 61 (6) (2009) 478–486.
[88] G. Gollavelli, Y.-C. Ling, Multi-functional graphene as an in vitro and in vivo imaging probe, Biomaterials 33 (8) (2012) 2532–2545.
[89] W. Wu, et al., Evaluation of the toxicity of graphene oxide exposure to the eye, Nanotoxicology 10 (9) (2016) 1329–1340.

CHAPTER 12

Targeted nanoparticles for treating infectious diseases

Viswanathan A. Aparna, Raja Biswas, R. Jayakumar
Amrita Centre for Nanosciences and Molecular Medicine, Amrita Institute of Medical Sciences and Research Centre, Amrita Vishwa Vidyapeetham, Kochi, India

Contents

12.1 Introduction

Infectious diseases caused by fungi, viruses, bacteria, and parasites are accountable for many deaths worldwide. According to WHO Global Health Estimates 2015, infectious and parasitic diseases account for 10% of the global mortality. Tuberculosis remained the most lethal infectious disease, causing 1.4 million deaths worldwide in 2015. The death toll of diarrheal diseases nearly halved between 2000 and 2015, but still caused 1.4 million deaths in 2015. Similarly, HIV/AIDS killed fewer people during the same period, having killed 1.1 million people in 2015 compared to 1.5 million in 2000. Among parasitic and vector diseases, malaria caused 0.4 million deaths which is nearly 8% of the total deaths by infectious and parasitic diseases [1].

Infectious diseases can be differentiated into emerging and reemerging infectious diseases. Emerging infectious diseases are due to infections which are newly identified, whereas reemerging infectious diseases are infections which are not new, but are resistant to drugs when they reappear, making them complicate to treat or manage [2–4].

Biomimetic Nanoengineered Materials for Advanced Drug Delivery
https://doi.org/10.1016/B978-0-12-814944-7.00012-6

For each disease and agent, immense attempts have been made considering the recognition of its molecular attributes and biomarkers, including antigens, toxins, peptides, or nucleic acids. Molecular medicine has enriched our insight of disease machinery and has fetched clinicians a collection of diagnostic tools that can abate the clinical dilemma [5–9]. Among these tools are nanosystems, which deliver ingenious molecular identification tools and bedside diagnostic solutions [10–13].

The 1970s witnessed the application of nanotechnology in efficient delivery of drugs for the treatment of different acute and chronic ailments [14, 15]. Various therapeutic molecules like proteins, drugs, nucleic acids, and peptides can be easily encapsulated or conjugated in nanomaterials enabling improved therapeutic efficiency. The encapsulation of biomolecules in nanoparticles ensure the protection of the molecules from degradation via proteases and DNases thereby ensuring sustainability and increased circulation time exclusive of any serious side effects [16].

Targeted delivery by means of nanoparticles can bestow enhanced intracellular drug concentration since the drug is delivered to a specific tissue or cell. Receptor-mediated endocytosis using suitable ligands is a feasible method. Specialized targeting strategies using nanocarriers including liposomes, dendrimers, nanoemulsions, polymeric nanoparticles, etc., are demonstrated for deadly diseases like malaria, HIV, tuberculosis, etc.

This chapter is an inclusive report of the current advancements in the targeted drug delivery of antimicrobial drugs using nanocarriers. The potential of nanocarriers for the treatment of various infectious diseases are reviewed in depth.

12.2 Infectious diseases

Infectious diseases are numerous times responsible for an erratic global impact like the pneumonic plague of the 14th century [17], influenza disease of 19th century [18], and the current most prominent one being HIV and AIDs [19]. Even though they have high potential of transmissibility, they can be prevented and eradicated as well. High replicative and mutational characteristics of pathogen make them easily adaptable, providing them with an evolutionary benefit over human host [20, 21].

Pathogens are organisms that instigate ailment or ill health to their host. Pathogens damage the host tissue during their process of invasion or by the secretion of toxins. Pathogens can be obligate or opportunistic. Obligate pathogens invade a healthy host and cause diseases whereas opportunistic pathogens are normally part of the normal flora but causes infection in immunocompromised host. In the process of invading host to obtain the necessary nutrients they need, pathogens cause host dysfunction. The process and mechanism of triggering host dysfunction is called pathogenesis. The ability of pathogens to trigger or produce infectious disease in their host is known as pathogenicity. A measure of degree of pathogenicity, extend/degree of mutilation instigated by pathogen, route of entry, host defences, and virulence factors is known as virulence. It is important to note that these pathogens do not want to harm the host on purpose, but purely in quest for an apt environment with the required nutrients (mainly iron) essential to survive and reproduce that is, pathogens exploit host-pathogen relationship to get the supply of metabolic materials/nutrients they need to survive and reproduce [22, 23]. Some pathogens seize from intracellular supplies (intracellular pathogens) and others grab from extracellular supplies (extracellular pathogens) which includes plasma with sugars, vitamins, minerals, etc.

Although the host have all the nutrients, the survival and growth of pathogen inside the host cell is challenging due to the host environmental conditions. The host extracellular and intracellular environment is inundated with immunosurveillant cells and components that impart the host immunity to foreign bodies. This protection force can be innate immune system (macrophages, natural killer cells, complements, cytokines, etc.) or the adaptive immune system namely B-cells and T-cells. Following the invasion of host by pathogens, they thrive inside the harsh environment of host by

(i) colonizing with the help of invasins and integrins/fibronectin interactions to prevent being washed away [24, 25];
(ii) finding suitable niche which can provide them with essential nutrients; and
(iii) outliving the innate and adaptive immune responses.

Extracellular microorganisms have established different practices to evade complement activation and phagocytosis mechanism of the host defence. Some microbes (e.g., *Streptococcus*) synthesize enzymes like c3b peptidase that cleaves c3b into inactive components inhibiting the complement pathway [26, 27]. *Salmonella* inhibits the membrane attack complex assembly using long smooth chain of lipopolysaccharide [28]. Some strains of *Staphylococcus aureus* dodge phagocytosis by producing leukocidins, an enzyme that kills phagocytes and some others bind to the F_c region of I_g or produce a slippery capsule to prevent opsonization [29–31].

Extracellular pathogens bring about host destruction mainly by the toxins they release. These toxins can be exotoxins or endotoxins. Exotoxins are released by bacteria into the environment and endotoxins are secreted by Gram negative bacteria and part of their cell wall. For example, *S. aureus* produce (exotoxin) superantigens that cause toxic shock syndrome and Gram negative bacteria release (endotoxin) lipopolysaccharide during their destruction [32, 33].

Intracellular pathogens replicate and grow by dodging the phagolysosomes using two mechanisms:

(i) Lysing or bypassing phagolysosomal vesicles and then inhabiting freely in the cytoplasm.

(ii) Surviving in phagosomes [34].

In the case of the survival of pathogens in phagosomes, either the fusion of phagosomes with lysosomes is impeded or the microbe will be resilient to degradative enzymes. *Mycobacterium tuberculosis* survives inside the phagocytes (macrophages) by arresting the phagosome biogenesis [35]. *Coxiella burnetii* converts phagosome into a spacious parasitophorous vacuole and flourish inside it [36]. *Neisseria gonorrhoeae* avoid neutrophil clearance in acute gonorrhoea by prolonging the fusion of primary granule and phagosome, thereby preventing phagolysosome formation [37]. *Legionella pneumophila* bring down host cell pathways to transform its growing phagosome into a *Legionella* containing vacuole, which functions as bacterial replication niche [38]. *Cryptococcus neoformans* are adapted to survive in acidic environments and therefore resides and replicates inside the phagolysosomal compartments [39]. *Yersinia pestis* prevents the acidification of vacuoles and proliferates inside the macrophages [40]. Extremely virulent *Toxoplasma gondii* resides in host cell vacuoles and resist typical phagosome-lysosome fusion [41].

Finally, there are medications or antimicrobials which are used to aid the host immune system to effectively overpower these pathogens. Extracellular pathogens are more manageable by antimicrobial treatment (except developing as biofilm) [42]. Management of intracellular pathogens poses some challenges. Antimicrobial agents should be conveyed into host cell and even this step meets challenges from cell local surroundings which can decrease their activity. Host tissue injury can also occur during selective toxicity and extermination of pathogen. This demands for the development of targeted drug delivery devices [43].

12.3 Effective antimicrobial activity of nanomaterials

Acidic antibiotics like β-lactamines during targeted delivery face poor uptake by the cells leading to the unsuccessful antimicrobial therapy. Basic antibiotics like aminoglycosides even if they successfully penetrate the cell membrane get inactivated in the acidic cytoplasm of the cells. Clindamycin, lincosamide, quinolones, etc., easily penetrates into the cell but faces rapid efflux from the cells leading to inefficacy of the drugs [44]. Superior surface area-to-volume ratio, distinctive physicochemical properties, and surface charges make nanomaterials suitable for antimicrobial action [45]. Nanoparticles enable delivery of drugs into the cell without getting inactivated during penetration. Unlike the drug molecules which diffuse into the cell, nanoparticles are taken up inside the cell by the process of endocytosis. The efficiency of nanoparticles for targeting drugs to the cells is depicted in Fig. 12.1.

Properties such as size, shape, surface functionalization, and composition can be tweaked to develop able-bodied nanoparticles for targeting purposes. Targeting nanoparticles generally have (i) coating to achieve a stealth property (to prevent the nanoparticles from being opsonized) or targeting, (ii) therapeutic molecule embedded inside the core or on the surface, (iii) fusogenic lipids which help in the release of drug at acidic microenvironment, and (iv) targeting ligand to deliver the drug to a specific cell [46, 47]. A general structure of targeted nanoparticle is illustrated in Fig. 12.2.

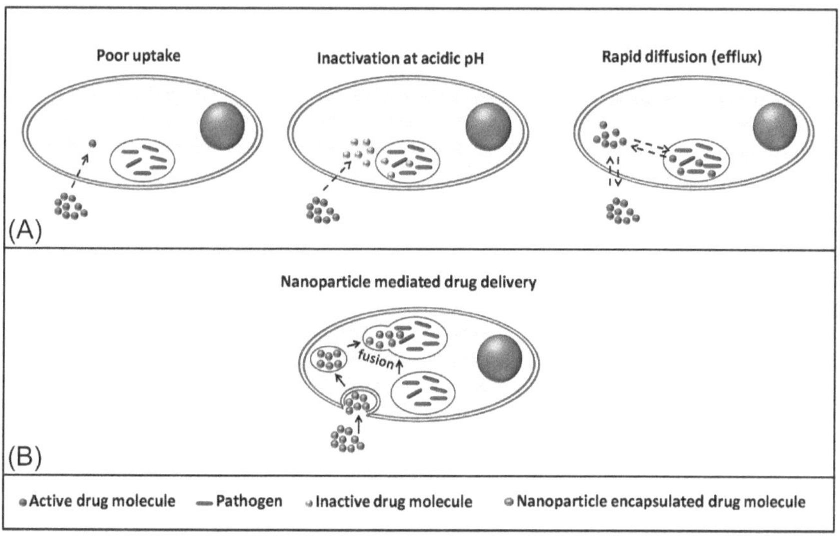

Fig. 12.1 Illustration of targeted drug delivery using nanoparticles for intracellular infections (A) the problems faced by bare drugs when they reach the cell and (B) nanoparticle-mediated drug delivery.

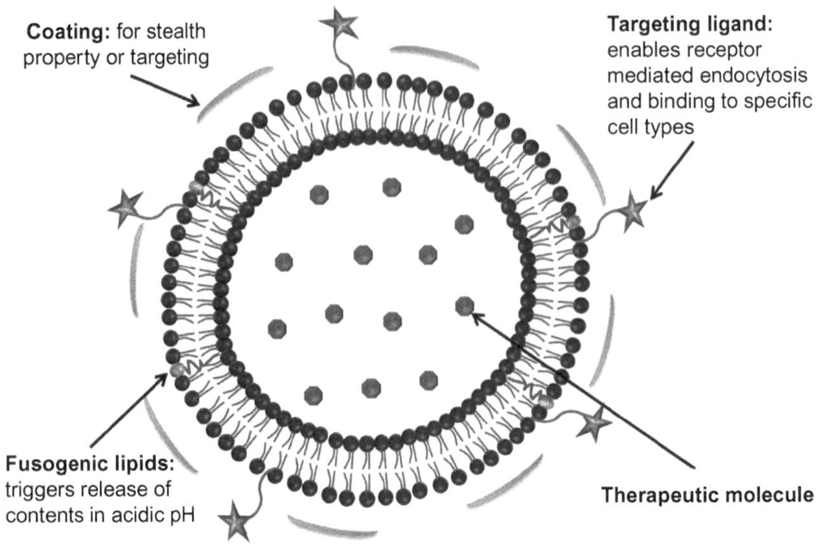

Fig. 12.2 Schematic representation of targeted nanoparticles with coating, targeting ligands, fusogenic lipids, and therapeutic molecules embedded inside them.

12.4 Targeted nanoparticles for infectious diseases

In order to improve the biocompatibility, targeting efficiency, and better dispensability in solution, the synthesized nanoparticles are surface functionalized using carbohydrates, proteins, or nucleic acids by the method of simple coating (physical method) or bioconjugation (chemical method) [48, 49]. Electrostatic and hydrophobic interactions between biomolecules and nanoparticles are employed in physical adsorption. However, physical adsorption has different drawbacks including uneven distribution, detachment, and different orientation of biomolecules attached on nanoparticles. In chemical modification, nanoparticles are functionalized using amine, sulfide, or carboxyl groups via covalent bond [50]. This method circumvents the physical adsorption inadequacies, and is therefore the better chosen mechanism for the attachment of biomolecules to nanoparticles. Nanoparticles surface functionalized with biomolecules have great applications such as pathogen detectors, anti-microbial as well as other diagnostic devices [49, 51–53].

12.5 Liposomes as drug carriers

Liposomes are bilayered lipid assembly in the form of a sphere, which imitates the cell membrane. This vesicle consists of amphiphilic lipid molecules and can promptly fuse with the pathogen, ensuing direct discharge of drugs to the core of the microbe. Liposomes can convey both hydrophilic and hydrophobic molecules (hydrophobic drugs are transported in its core, whereas hydrophilic molecules are transported inside the lipid molecules in the membrane). Liposomes can be easily functionalized with stealth material to dodge the macrophagic uptake [54].

12.6 Polymeric nanoparticles as drug carriers

Some exclusive features of polymeric nanoparticles as carriers for antimicrobial treatment are as follows:

- structural stability
- nanoparticles can be developed with a narrow size range
- ability to control drug release as well as other characteristics by altering its chemistry
- high modification potential so as to incorporate ligands to its surface

In many antimicrobial therapies, the polymeric nanoparticles are surface functionalized with lectin, which enable the nanoparticles to bind with the carbohydrates present on the bacterial cell walls release its contents to the interior of the bacteria, resulting in the death of the pathogen [55].

Amphotericin B in *O*-palmitoyl mannan-coated emulsions could successfully eliminate intracellular amastigotes of *Leishmania donovani* residing in macrophages [56]. The same drug when encapsulated in tuftsin-bearing liposomes effectively inhibited *Candida albicans* and *L. donovani* [57, 58]. Mannose conjugated gelatine nanoparticles, lectinized lipo-polymerosome, and poly(lactic acid-co-glycolic acid-polyethylene glycol) nanoparticles were also used to target and efficiently kill *L. donovani* [59–61]. In vitro targeted antifungal activity of amphotericin B against *Candida glabrata* was enhanced when amphotericin B was encapsulated in sulfated chitosan nanoparticles and carboxymethyl-ι-carrageenan conjugated gelatin nanoparticles [62, 63].

Targeted delivery of azidothymidine showed efficient treatment of HIV/AIDs. Azidothymidine was encapsulated in hexylcyanoacrylate nanoparticles and acetylated LDL microemulsion for targeting purposes [64–66]. PLGA nanoparticles and poly(amidoamine) dendrimers were used as carriers for the drug azithromycin, a macrolide antibiotic for the management of *Chlamydia trachomatis* infections [67, 68]. Ceftriaxone sodium, a β-lactam, third-generation cephalosporin antibiotic, showed better intracellular inhibition of *Salmonella,* when is delivered to the cell inside the chitosan nanoparticles [69].

Ciprofloxacin was delivered inside fucoidan-coated chitosan nanoparticles and mannosylated liposomes for better management of *Salmonella* and other intracellular parasites [70, 71]. Didanosine, indinavir, and stavudine are encapsulated, respectively, in mannan-coated gelatin nanoparticles, immunoliposomes, and mannosylated liposomes for the management of HIV/AIDS [72–74]. *Brucella melitensis* and *Salmonella enterica* were efficiently eradicated from the intracellular compartments by the method of targeting gentamicin using silica, PLGA nanoparticles, and magnetite block ionomer complexes [75–79].

Hamycin, a polyene antimycotic, when contained in neoglycoprotein conjugated liposomes, exhibited better inhibition of *L. donovani* [80]. Improved anti-*Mycobacterium tuberculosis* activity was demonstrated when rifampin and rifabutin was nanoformulated using tuftsin-bearing liposomes, poly(ethyleneimine)-coated mesoporous silica nanoparticle, and mannosylated solid lipid nanoparticles [81–83]. Rifampicin-mannose-conjugated chitosan nanoparticle formulation showed superior anti-leishmanial activity compared to bare rifampicin [84]. Methicillin-resistant *S. aureus* (MRSA) was effectively killed when vancomycin or teicoplanin were encapsulated in phosphatidyl choline, diacetyl phosphate, and cholesterol nanoparticles [85].

Table 12.1 illustrates different types of polymeric nanoparticles used for targeted delivery of drugs for treatment of infectious diseases. Elbi S et al. have used fucoidan to target chitosan nanoparticles (CNPs) to *Salmonella* infected macrophages. Fucoidan was coated on the surface of chitosan nanoparticles (Fu-CNPs) inorder to increase its uptake by macrophage cells [71]. They have demonstrated the above by using fluorescence microscopy (Fig. 12.3A), where FITC loaded Fu-CNPs had better uptake than FITC loaded chitosan nanoparticles (CNPs) by RAW 264.7 macrophage cells. They have also shown enhanced intracellular *Salmonella* death by ciprofloxacin (c)-containing Fu-cCNPs in comparison to cCNPs and bare ciprofloxacin (Fig. 12.3B and C).

12.7 Current scenario

Despite numerous studies in the region of targeted antimicrobial delivery, only a small number of formulations reached clinical-preclinical trials and a very few got marketed. Amphotericin B formulations of nanoemulsions and carbon nanotubes are at the preclinical stage, whereas its liposomal formulation is currently marketed [86–89]. Another lipid-based non-liposomal formulation of Amphotericin B is marketed outside the

Table 12.1 List of different nanoparticle formulations studied for the treatment of intracellular infections.

Therapeutic molecule	Formulation type	Disease condition/ pathogen which is inhibited	References
Amphotericin B	*O*-Palmitoyl mannan-coated emulsions	Visceral leishmaniasis	Gupta et al. [56]
Amphotericin B	Tuftsin-bearing liposomes	*Candida albicans*	Khan and Owais [58]
Amphotericin B	Tuftsin-bearing liposomes	*Leishmania donovani*	Agrawal et al. [57]
Amphotericin B	Mannose conjugated gelatin nanoparticles	Visceral leishmaniasis	Nahar et al. [61]
Amphotericin B	Lectinized lipo-polymerosome	Visceral leishmaniasis	Gupta et al. [59]
Amphotericin B	Poly(lactic acid-co-glycolic acid-polyethylene glycol) nanoparticles	*L. donovani*	Kumar et al. [60]
Amphotericin B	Sulfated chitosan nanoparticles	*Candida glabrata*	Sandhya et al. [63]
Amphotericin B	Carboxymethyl-ι-carrageenan conjugated gelatin nanoparticles	*Candida glabrata*	Aparna et al. [62]
Azidothymidine	Hexylcyanoacrylate nanoparticles	HIV/AIDS	Löbenberg et al. [65], Löbenberg and Kreuter [66]
Azidothymidine	Acetylated LDL microemulsion	HIV/AIDS	Hu et al. [64]
Azithromycin/ rifampin	PLGA nanoparticles	*Chlamydia trachomatis*	Toti et al. [68]
Azithromycin	Poly(amidoamine) dendrimers	*C. trachomatis*	Mishra et al. [67]
Ceftriaxone Sodium	Chitosan nanoparticles	*Salmonella Typhimurium*	Zaki and Hafez [69]
Ciprofloxacin	Mannosylated liposomes	Intracellular parasitic infections	Chono et al. [70]
Ciprofloxacin	Fucoidan-coated chitosan nanoparticles	*Salmonella*	Elbi et al. [71]
Didanosine	Mannan-coated gelatin nanoparticles	HIV/AIDS	Kaur et al. [74]
Gentamicin	Silica nanoparticles	*Salmonella enterica*	Seleem et al. [79]

Continued

Table 12.1 List of different nanoparticle formulations studied for the treatment of intracellular infections—cont'd

Therapeutic molecule	Formulation type	Disease condition/ pathogen which is inhibited	References
Gentamicin	PLGA nanoparticles	*Brucella melitensis*	Imbuluzqueta et al. [75]
Gentamicin	Magnetite block ionomer complexes	*Brucella melitensis*	Jain-Gupta et al. [76]
Gentamicin	PLGA nanoparticles	*Brucella melitensis*	Lecaroz et al. [78]
Gentamicin	PLGA nanoparticles	*Brucella melitensis*	Lecároz et al. [77]
Hamycin	Neoglycoprotein conjugated liposomes	*L. donovani*	Kole et al. [80]
Indinavir	Immunoliposomes	HIV/AIDS	Gagneé et al. [72]
Rifampin	Tuftsin-bearing liposomes	*Mycobacterium tuberculosis*	Agarwal et al. [81]
Rifampicin	Mannose-conjugated chitosan nanoparticles	Visceral leishmaniasis	Chaubey and Mishra [84]
Rifampin/ isoniazid	Poly(ethyleneimine)-coated mesoporous silica nanoparticle	*Mycobacterium tuberculosis*	Hwang et al. [82]
Rifabutin	Mannosylated solid lipid nanoparticles	*Mycobacterium tuberculosis*	Nimje et al. [83]
Stavudine	Mannosylated liposomes	HIV	Garg et al. [73]
Vancomycin or teicoplanin	Phosphatidyl choline, diacetyl phosphate, and cholesterol nanoparticles	Methicillin-resistant *Staphylococcus aureus*	Onyeji et al. [85]

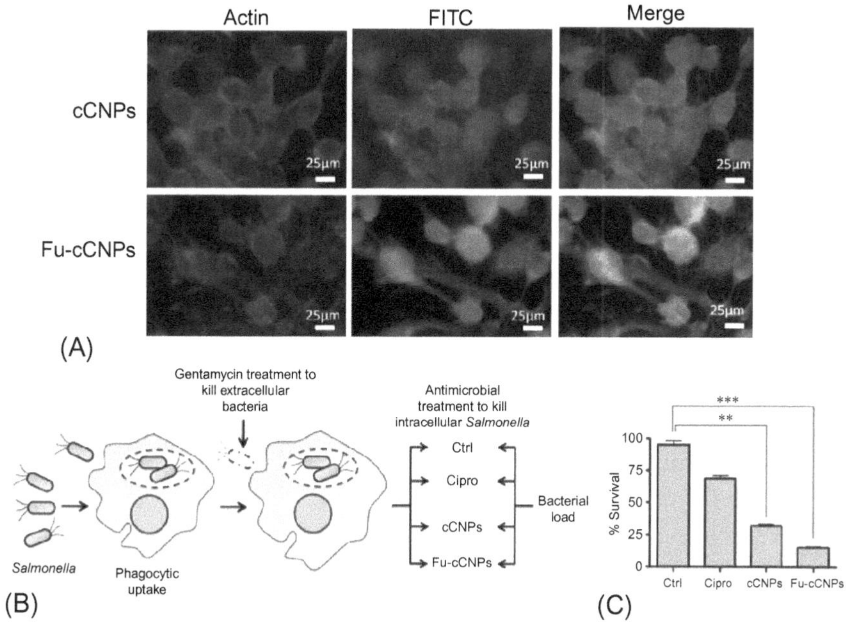

Fig. 12.3 (A) Macrophage cells were stained with phalloidin dye to visualize actin filaments (*red—dark gray* in the print version). FITC was encapsulated into the nanoparticle-Fu-cCNPs and cCNPs (*green—light gray* in the print version) and the uptake of nanoparticles by the cells were visualized using fluorescence microscopy. The last image shows the merged photograph of cells showing both nanoparticles and actin filaments. (B) Schematic representation of intracellular survival assay of *Salmonella* in RAW 264.7 macrophage cells. (C) Bar diagram showing the reduced intracellular count of Salmonella when treated with ciprofloxacin-containing fucoidan coated chitosan nanoparticles. *(Reprinted from S. Elbi, T.R. Nimal, V.K. Rajan, G. Baranwal, R. Biswas, R. Jayakumar, S. Sathianarayanan, Fucoidan coated ciprofloxacin loaded chitosan nanoparticles for the treatment of intracellular and biofilm infections of* Salmonella. *Colloids Surf. B: Biointerfaces 160 (2017) 40–47, Copyright (2017), with permission from Elsevier).*

United States, which is suggested for the management of systemic fungal infections. The drug is postulated to reach fungal infected cells with least uptake by normal human cells [90, 91]. Benzathine penicillin G in nanoemulsions completed preclinical evaluations [92].

The significant in the midst of these is amphotericin B liposomal formulation. It has been approved and recommended for the management of candidiasis, aspergillosis, and cryptococcosis by the FDA. It (AmBisome; Gilead Sciences, USA) has also been certified for anti-Visceral leishmaniasis therapy in numerous European countries and the United States.

12.8 Outlook

Regardless of the immense advancements in nanoparticle-mediated delivery of therapeutic molecules, numerous challenges are still unacknowledged. Primarily, the treatment of infections by drug-resistant bacteria persists as a foremost challenge. Combinational therapy can be a solution to this issue, since they are expected to have a synergistic effect on antimicrobial action and together can overthrow the pathogen defence methods. The efficacy of combinational therapy in clinics is still a question. Secondly, the mechanism of how the drug encapsulated in nanoparticle can distinguish between the pathogen inhabited cell and uninhabited cell needs to be studied. Targeting specificity should be improved further by considering the above, which will help in further reduction of drug dosage and related toxicity issues. Thirdly, studies on how the endocytosed nanoparticles reach the pathogens residing inside other vacuoles in the same host cell are limited. The efficacy of the formulation can be enhanced by designing nanoparticles which can adhere to the pathogen surface until most of the drug is released. Lastly, researches on targeting of nanoparticles to unprofessional phagocytes are restricted. All these challenges can be addressed by combined interdisciplinary involvement of pharmacists, nanoengineers, biotechnologists, chemical engineers, and microbiologists.

12.9 Conclusions

In short, majority of antimicrobial medications are difficult to administer owing to their poor water solubility, high cytotoxicity as well as their quick degradation and clearance from the blood stream. The microbicidal activities of these drugs against intracellular pathogens are also restricted due to reduced membrane penetration. Studies have reported effective antimicrobial drug delivery to infection sites with the help of different types of nanoparticles. Although majority of antimicrobial drug delivery using nanoparticles are currently in preclinical stages, some have been permitted for medical treatments in clinical scenarios. With the ongoing attempts in this platform, there is negative uncertainty that drug delivery using nanoparticles will keep on enhance the treatment modalities.

References

[1] WHO. Global Health Estimates 2015: Disease Burden by Cause, Age, Sex, by Country and by Region. WHO (Vol. Multimedia). Retrieved from http://www.who.int/healthinfo/global_burden_disease/estimates/en/index2.html, 2017.

[2] A.S. Fauci, D.M. Morens, The perpetual challenge of infectious diseases, N. Engl. J. Med. 366 (5) (2012) 454–461.

[3] D.M. Morens, A.S. Fauci, Emerging infectious diseases: threats to human health and global stability, PLoS Pathog. 9 (7) (2013) e1003467.
[4] C.R. Parrish, et al., Cross-species virus transmission and the emergence of new epidemic diseases, Microbiol. Mol. Biol. Rev. 72 (3) (2008) 457–470.
[5] F. Blasi, D. Stolz, F. Piffer, Biomarkers in lower respiratory tract infections, Pulm. Pharmacol. Ther. 23 (6) (2010) 501–507.
[6] N. Chauhan, S. Tiwari, U. Jain, Potential biomarkers for effective screening of neonatal sepsis infections: an overview, Microb. Pathog. (2017), https://doi.org/10.1016/j.micpath.2017.03.042.
[7] B. Müller, P. Schuetz, A. Trampuz, Circulating biomarkers as surrogates for bloodstream infections, Int. J. Antimicrob. Agents 30 (Suppl. 1) (2007) 16–23.
[8] P. Schuetz, M. Christ-Crain, B. Müller, Biomarkers to improve diagnostic and prognostic accuracy in systemic infections, Curr. Opin. Crit. Care 13 (5) (2007) 578–585.
[9] G. Walzl, K. Ronacher, W. Hanekom, T.J. Scriba, A. Zumla, Immunological biomarkers of tuberculosis, Nat. Rev. Immunol. 11 (5) (2011) 343–354.
[10] F. Degliangeli, P.P. Pompa, R. Fiammengo, Nanotechnology-based strategies for the detection and quantification of microRNA, Chem. Eur. J. 20 (31) (2014) 9476–9492.
[11] J.D. Driskell, R.A. Tripp, Emerging technologies in nanotechnology-based pathogen detection, Clin. Microbiol. Newsl. 31 (18) (2009) 137–144.
[12] C. Kaittanis, S. Santra, J.M. Perez, Emerging nanotechnology-based strategies for the identification of microbial pathogenesis, Adv. Drug Deliv. Rev. 62 (4–5) (2010) 408–423.
[13] M.A. Syed, S.H.A. Bokhari, Gold nanoparticle based microbial detection and identification, J. Biomed. Nanotechnol. 7 (2) (2011) 229–237.
[14] G. Gregoriadis, Drug entrapment in liposomes, FEBS Lett. 36 (3) (1973) 292–296.
[15] P.A. Kramer, Albumin microspheres as vehicles for achieving specificity in drug delivery, J. Pharm. Sci. 63 (10) (1974) 1646–1647.
[16] J.R. Kanwar, R.R. Mohan, R.K. Kanwar, K. Roy, R. Bawa, Applications of aptamers in nanodelivery systems in cancer, eye and inflammatory diseases, Nanomedicine 5 (9) (2010) 1435–1445.
[17] D.M. Morens, G.K. Folkers, A.S. Fauci, Emerging infections: a perpetual challenge, Lancet Infect. Dis. 8 (11) (2008) 710–719.
[18] D.M. Morens, A.S. Fauci, The 1918 influenza pandemic: insights for the 21st Century, J. Infect. Dis. 195 (7) (2007) 1018–1028.
[19] K.M. De Cock, H.W. Jaffe, J.W. Curran, Reflections on 30 years of AIDS, Emerg. Infect. Dis. 17 (6) (2011) 1044–1048.
[20] J. Lederberg, Infectious history, Science 288 (5464) (2000) 287–293.
[21] D.M. Morens, G.K. Folkers, A.S. Fauci, The challenge of emerging and re-emerging infectious diseases, Nature 430 (6996) (2004) 242–249.
[22] M. Caza, J.W. Kronstad, Shared and distinct mechanisms of iron acquisition by bacterial and fungal pathogens of humans, Front. Cell. Infect. Microbiol. 3 (2013) 80.
[23] K.G. Wooldridge, P.H. Williams, Iron uptake mechanisms of pathogenic bacteria, FEMS Microbiol. Rev. 12 (4) (1993) 325–348.
[24] H.H. Niemann, W.D. Schubert, D.W. Heinz, Adhesins and invasins of pathogenic bacteria: a structural view, Microbes Infect. 6 (1) (2004) 101–112.
[25] A. Scibelli, S. Roperto, L. Manna, L.M. Pavone, S. Tafuri, R. Della Morte, N. Staiano, Engagement of integrins as a cellular route of invasion by bacterial pathogens, Vet. J. 173 (3) (2007) 482–491.
[26] G. Agrahari, Z. Liang, K. Glinton, S.W. Lee, V.A. Ploplis, F.J. Castellino, *Streptococcus pyogenes* employs strain-dependent mechanisms of C3b inactivation to inhibit phagocytosis and killing of bacteria, J. Biol. Chem. 291 (17) (2016) 9181–9189.

[27] F. Goldblatt, J. Yuste, D.A. Isenberg, A. Rahman, J. Brown, Impaired C3b/iC3b deposition on *Streptococcus pneumoniae* in serum from patients with systemic lupus erythematosus, Rheumatology 48 (12) (2009) 1498–1501.
[28] C.W. Sheu, E. Freese, Lipopolysaccharide layer protection of gram-negative bacteria against inhibition by long-chain fatty acids, J. Bacteriol. 115 (3) (1973) 869–875.
[29] S.D. Kobayashi, F.R. DeLeo, *Staphylococcus aureus* protein A promotes immune suppression, MBio 4 (5) (2013) e00764-13.
[30] K. O'Riordan, J.C. Lee, *Staphylococcus aureus* capsular polysaccharides, Clin. Microbiol. Rev. 17 (1) (2004) 218–234.
[31] F. Vandenesch, G. Lina, T. Henry, *Staphylococcus aureus* hemolysins, bi-component leukocidins, and cytolytic peptides: a redundant arsenal of membrane-damaging virulence factors? Front. Cell. Infect. Microbiol. 2 (2012) 12.
[32] J.K. McCormick, J.M. Yarwood, P.M. Schlievert, Toxic shock syndrome and bacterial superantigens: an update, Annu. Rev. Microbiol. 55 (1) (2001) 77–104.
[33] X. Wang, P.J. Quinn, Endotoxins: lipopolysaccharides of gram-negative bacteria, Subcell. Biochem. 53 (2010) 3–25.
[34] R.S. Flannagan, G. Cosío, S. Grinstein, Antimicrobial mechanisms of phagocytes and bacterial evasion strategies, Nat. Rev. Microbiol. 7 (5) (2009) 355–366.
[35] V. Deretic, et al., *Mycobacterium tuberculosis* inhibition of phagolysosome biogenesis and autophagy as a host defence mechanism, Cell. Microbiol. 8 (5) (2006) 719–727.
[36] D.E. Voth, R.A. Heinzen, Lounging in a lysosome: the intracellular lifestyle of *Coxiella burnetii*, Cell. Microbiol. 9 (4) (2007) 829–840.
[37] M.B. Johnson, A.K. Criss, *Neisseria gonorrhoeae* phagosomes delay fusion with primary granules to enhance bacterial survival inside human neutrophils, Cell. Microbiol. 15 (8) (2013) 1323–1340.
[38] A. Prashar, M.R. Terebiznik, *Legionella pneumophila*: homeward bound away from the phagosome, Curr. Opin. Microbiol. 23 (2015) 86–93.
[39] S.M. Levitz, S.H. Nong, K.F. Seetoo, T.S. Harrison, R.A. Speizer, E.R. Simons, *Cryptococcus neoformans* resides in an acidic phagolysosome of human macrophages, Infect. Immun. 67 (2) (1999) 885–890.
[40] C. Pujol, K.A. Klein, G.A. Romanov, L.E. Palmer, C. Cirota, Z. Zhao, J.B. Bliska, *Yersinia pestis* can reside in autophagosomes and avoid xenophagy in murine macrophages by preventing vacuole acidification, Infect. Immun. 77 (6) (2009) 2251–2261.
[41] L.D. Sibley, E. Weidner, J.L. Krahenbuhl, Phagosome acidification blocked by intracellular *Toxoplasma gondii*, Nature 315 (6018) (1985) 416–419.
[42] I. Olsen, Biofilm-specific antibiotic tolerance and resistance, Eur. J. Clin. Microbiol. Infect. Dis. 34 (5) (2015) 877–886.
[43] A. Ranjan, N. Pothayee, M.N. Seleem, S.M. Boyle, R. Kasimanickam, J.S. Riffle, N. Sriranganathan, Nanomedicine for intracellular therapy, FEMS Microbiol. Lett. 332 (1) (2012) 1–9.
[44] H. Pinto-Alphandary, A. Andremont, P. Couvreur, Targeted delivery of antibiotics using liposomes and nanoparticles: research and applications, Int. J. Antimicrob. Agents 13 (3) (2000) 155–168.
[45] E. Weir, A. Lawlor, A. Whelan, F. Regan, The use of nanoparticles in anti-microbial materials and their characterization, Analyst 133 (7) (2008) 835.
[46] H. Hatakeyama, E. Ito, H. Akita, M. Oishi, Y. Nagasaki, S. Futaki, H. Harashima, A pH-sensitive fusogenic peptide facilitates endosomal escape and greatly enhances the gene silencing of siRNA-containing nanoparticles *in vitro* and *in vivo*, J. Control. Release 139 (2) (2009) 127–132.
[47] S.M. Moghimi, J. Szebeni, Stealth liposomes and long circulating nanoparticles: critical issues in pharmacokinetics, opsonization and protein-binding properties, Prog. Lipid Res. 42 (6) (2003) 463–478.

[48] A.I. Lopez, A. Kumar, M.R. Planas, Y. Li, T.V. Nguyen, C. Cai, Biofunctionalization of silicone polymers using poly(amidoamine) dendrimers and a mannose derivative for prolonged interference against pathogen colonization, Biomaterials 32 (19) (2011) 4336–4346.
[49] M. Veerapandian, K. Yun, Functionalization of biomolecules on nanoparticles: specialized for antibacterial applications, Appl. Microbiol. Biotechnol. 90 (5) (2011) 1655–1667.
[50] R. Mout, D.F. Moyano, S. Rana, V.M. Rotello, Surface functionalization of nanoparticles for nanomedicine, Chem. Soc. Rev. 41 (7) (2012) 2539.
[51] L. Qi, Z. Xu, X. Jiang, C. Hu, X. Zou, Preparation and antibacterial activity of chitosan nanoparticles, Carbohydr. Res. 339 (16) (2004) 2693–2700.
[52] M. Rai, A. Yadav, A. Gade, Silver nanoparticles as a new generation of antimicrobials, Biotechnol. Adv. 27 (1) (2009) 76–83.
[53] C.M. Yam, C.M. Pradier, M. Salmain, N. Fischer-Durand, G. Jaouen, Molecular recognition of avidin on biotin-functionalized gold surfaces detected by FT-IRRAS and use of metal carbonyl probes, J. Colloid Interface Sci. 245 (1) (2002) 204–207.
[54] A. Akbarzadeh, et al., Liposome: classification, preparation, and applications, Nanoscale Res. Lett. 8 (1) (2013) 1–8.
[55] R.B. Umamaheshwari, N.K. Jain, Receptor mediated targeting of lectin conjugated gliadin nanoparticles in the treatment of *Helicobacter pylori*, J. Drug Target. 11 (7) (2003) 415–424.
[56] S. Gupta, A. Dube, S.P. Vyas, Antileishmanial efficacy of amphotericin B bearing emulsomes against experimental visceral leishmaniasis, J. Drug Target. 15 (6) (2007) 437–444.
[57] A.K. Agrawal, A. Agrawal, A. Pal, P.Y. Guru, C.M. Gupta, Superior chemotherapeutic efficacy of amphotericin B in tuftsin-bearing liposomes against *Leishmania donovani* infection in hamsters, J. Drug Target. 10 (1) (2002) 41–45.
[58] M.A. Khan, M. Owais, Toxicity, stability and pharmacokinetics of amphotericin B in immunomodulator tuftsin-bearing liposomes in a murine model, J. Antimicrob. Chemother. 58 (1) (2006) 125–132.
[59] P.K. Gupta, et al., Exploitation of lectinized lipo-polymerosome encapsulated amphotericin b to target macrophages for effective chemotherapy of visceral leishmaniasis, Bioconjug. Chem. 25 (6) (2014) 1091–1102.
[60] R. Kumar, G.C. Sahoo, K. Pandey, V. Das, P. Das, Study the effects of PLGA-PEG encapsulated Amphotericin B nanoparticle drug delivery system against *Leishmania donovani*, Drug Deliv. 22 (3) (2015) 383–388.
[61] M. Nahar, V. Dubey, D. Mishra, P.K. Mishra, A. Dube, N.K. Jain, *In vitro* evaluation of surface functionalized gelatin nanoparticles for macrophage targeting in the therapy of visceral leishmaniasis, J. Drug Target. 18 (2) (2010) 93–105.
[62] V. Aparna, A.R. Melge, V.K. Rajan, R. Biswas, R. Jayakumar, C. Gopi Mohan, Carboxymethylated ι-carrageenan conjugated amphotericin B loaded gelatin nanoparticles for treating intracellular *Candida glabrata* infections, Int. J. Biol. Macromol. (2017) https://doi.org/10.1016/j.ijbiomac.2017.11.126.
[63] S. Murali, V. Aparna, K. Suresh, R. Biswas, R. Jayakumar, S. Sathianarayanan, Amphotericin B loaded sulphonated chitosan nanoparticles for targeting macrophages to treat intracellular *Candida glabrata* infections. Int. J. Biol. Macromol. (2018), https://doi.org/10.1016/j.ijbiomac.2018.01.028.
[64] J. Hu, H. Liu, L. Wang, Enhanced delivery of AZT to macrophages via acetylated LDL, J. Control. Release 69 (3) (2000) 327–335.
[65] R. Löbenberg, L. Araujo, J. Kreuter, Body distribution of azidothymidine bound to nanoparticles after oral administration, Eur. J. Pharm. Biopharm. 44 (2) (1997) 127–132.

[66] R. Löbenberg, J. Kreuter, Macrophage targeting of azidothymidine: a promising strategy for AIDS therapy, AIDS Res. Hum. Retrovir. 12 (18) (1996) 1709–1715.
[67] M.K. Mishra, et al., PAMAM dendrimer-azithromycin conjugate nanodevices for the treatment of *Chlamydia trachomatis* infections, Nanomedicine 7 (6) (2011) 935–994.
[68] U.S. Toti, B.R. Guru, M. Hali, C.M. McPharlin, S.M. Wykes, J. Panyam, J. A. Whittum-Hudson, Targeted delivery of antibiotics to intracellular chlamydial infections using PLGA nanoparticles, Biomaterials 32 (27) (2011) 6606–6613.
[69] N.M. Zaki, M.M. Hafez, Enhanced antibacterial effect of ceftriaxone sodium-loaded chitosan nanoparticles against intracellular *Salmonella typhimurium*, AAPS PharmSciTech 13 (2) (2012) 411–421.
[70] S. Chono, T. Tanino, T. Seki, K. Morimoto, Efficient drug targeting to rat alveolar macrophages by pulmonary administration of ciprofloxacin incorporated into mannosylated liposomes for treatment of respiratory intracellular parasitic infections, J. Control. Release 127 (1) (2008) 50–58.
[71] S. Elbi, T.R. Nimal, V.K. Rajan, G. Baranwal, R. Biswas, R. Jayakumar, S. Sathianarayanan, Fucoidan coated ciprofloxacin loaded chitosan nanoparticles for the treatment of intracellular and biofilm infections of *Salmonella*, Colloids Surf. B: Biointerfaces 160 (2017) 40–47.
[72] J.F. Gagneé, A. Deésormeaux, S. Perron, M.J. Tremblay, M.G. Bergeron, Targeted delivery of indinavir to HIV-1 primary reservoirs with immunoliposomes, Biochim. Biophys. Acta Biomembr. 1558 (2) (2002) 198–210.
[73] M. Garg, A. Asthana, H.B. Agashe, G.P. Agrawal, N.K. Jain, Stavudine-loaded mannosylated liposomes: in-vitro anti-HIV-I activity, tissue distribution and pharmacokinetics, J. Pharm. Pharmacol. 58 (5) (2006) 605–616.
[74] A. Kaur, S. Jain, A.K. Tiwary, Mannan-coated gelatin nanoparticles for sustained and targeted delivery of didanosine: *in vitro* and *in vivo* evaluation, Acta Pharma. 58 (1) (2008) 61–74.
[75] E. Imbuluzqueta, et al., Hydrophobic gentamicin-loaded nanoparticles are effective against *Brucella melitensis* infection in mice, Antimicrob. Agents Chemother. 57 (7) (2013) 3326–3333.
[76] N. Jain-Gupta, et al., Efficacies of gentamicin-loaded magnetite block ionomer complexes against chronic *Brucella melitensis* infection, J. Nanopart. Res. 15 (11) (2013) 2024.
[77] C. Lecároz, M.J. Blanco-Prieto, M.A. Burrell, C. Gamazo, Intracellular killing of *Brucella melitensis* in human macrophages with microsphere-encapsulated gentamicin, J. Antimicrob. Chemother. 58 (3) (2006) 549–556.
[78] M.C. Lecaroz, M.J. Blanco-Prieto, M.A. Campanero, H. Salman, C. Gamazo, Poly(D, L-lactide-coglycolide) particles containing gentamicin: pharmacokinetics and pharmacodynamics in *Brucella melitensis* infected mice, Antimicrob. Agents Chemother. 51 (4) (2007) 1185–1190.
[79] M.N. Seleem, P. Munusamy, A. Ranjan, H. Alqublan, G. Pickrell, N. Sriranganathan, Silica-antibiotic hybrid nanoparticles for targeting intracellular pathogens, Antimicrob. Agents Chemother. 53 (10) (2009) 4270–4274.
[80] L. Kole, K. Sarkar, S.B. Mahato, P.K. Das, Neoglycoprotein conjugated liposomes as macrophage specific drug carrier in the therapy of leishmaniasis, Biochem. Biophys. Res. Commun. 200 (1) (1994) 351–358.
[81] A. Agarwal, H. Kandpal, H.P. Gupta, N.B. Singh, C.M. Gupta, Tuftsin-bearing liposomes as rifampin vehicles in treatment of tuberculosis in mice, Antimicrob. Agents Chemother. 38 (3) (1994) 588–593.
[82] A.A. Hwang, B.Y. Lee, D.L. Clemens, B.J. Dillon, J.I. Zink, M.A. Horwitz, Tuberculosis: pH-responsive isoniazid-loaded nanoparticles markedly improve tuberculosis treatment in mice (Small 38/2015), Small 11 (38) (2015) 5065.

[83] N. Nimje, A. Agarwal, G.K. Saraogi, N. Lariya, G. Rai, H. Agrawal, G.P. Agrawal, Mannosylated nanoparticulate carriers of rifabutin for alveolar targeting, J. Drug Target. 17 (10) (2009) 777–787.

[84] P. Chaubey, B. Mishra, Mannose-conjugated chitosan nanoparticles loaded with rifampicin for the treatment of visceral leishmaniasis, Carbohydr. Polym. 101 (2014) 1101–1108.

[85] C.O. Onyeji, C.H. Nightingale, M.N. Marangos, Enhanced killing of methicillin-resistant *Staphylococcus aureus* in human macrophages by liposome-entrapped vancomycin and teicoplanin, Infection 22 (5) (1994) 338–342.

[86] B. Brime, Comparative pharmacokinetics and safety of a novel lyophilized amphotericin B lecithin-based oil-water microemulsion and amphotericin B deoxycholate in animal models, J. Antimicrob. Chemother. 52 (1) (2003) 103–109.

[87] R.N. Davidson, S.L. Croft, A. Scott, M. Maini, A.H. Moody, A.D. Bryceson, Liposomal amphotericin B in drug-resistant visceral leishmaniasis, Lancet 337 (8749) (1991) 1061–1062.

[88] V.K. Prajapati, K. Awasthi, S. Gautam, T.P. Yadav, M. Rai, O.N. Srivastava, S. Sundar, Targeted killing of *Leishmania donovani in vivo* and *in vitro* with amphotericin B attached to functionalized carbon nanotubes, J. Antimicrob. Chemother. 66 (4) (2011) 874–879.

[89] S. Sundar, J. Chakravarty, D. Agarwal, M. Rai, H.W. Murray, Single-dose liposomal amphotericin B for visceral leishmaniasis in India, N. Engl. J. Med. 362 (6) (2010) 504–512.

[90] S. De Marie, R. Jankengt, I.A.J.M. Bakker-Woudenberg, Clinical use of liposomal and lipid-complexed amphotericin B, J. Antimicrob. Chemother. 33 (5) (1994) 907–916.

[91] R. Janknegt, S. de Marie, I.A.J.M. Bakker-Woudenberg, D.J.A. Crommelin, Liposomal and lipid formulations of amphotericin B: clinical pharmacokinetics, Clin. Pharmacokinet. 23 (4) (1992) 279–291.

[92] N.S. Santos-Magalhães, A. Pontes, V.M.W. Pereira, M.N.P. Caetano, Colloidal carriers for benzathine penicillin G: nanoemulsions and nanocapsules, Int. J. Pharm. 208 (1–2) (2000) 71–80.

Index

Note: Page numbers followed by *f* indicate figures and *t* indicate tables.

E

F

G

H

I

J

L

M

FSC
www.fsc.org
MIX
Papier | Fördert
gute Waldnutzung
FSC® C083411

Kolibri 360 GmbH
Ferdinand-Jühlke-Straße 7
99095 Erfurt, Deutschland
produktsicherheit@kolibri360.de